EXPERIMENTAL ANIMAL PHYSIOLOGY AND BIOCHEMISTRY

AF651427

EXPERIMENTAL ANIMAL PHYSIOLOGY AND BIOCHEMISTRY

S C NIGAM
Formerly Professor of Zoology
Department of Zoology
Lucknow University, Lucknow

OMKAR
Reader
University of Gorakhpur
Gorakhpur, India

Publishing Globally

NEW AGE INTERNATIONAL (P) LIMITED, PUBLISHERS
LONDON • NEW DELHI • NAIROBI
Bangalore • Chennai • Cochin • Guwahati • Hyderabad • Kolkata • Lucknow • Mumbai
Visit us at **www.newagepublishers.com**

Copyright © 2003, New Age International (P) Ltd., Publishers
Published by New Age International (P) Ltd., Publishers
First Edition: 2003
Reprint: 2017

All rights reserved.
No part of this book may be reproduced in any form, by photostat, microfilm, xerography, or any other means, or incorporated into any information retrieval system, electronic or mechanical, without the written permission of the publisher.

GLOBAL OFFICES

- **New Delhi** NEW AGE INTERNATIONAL (P) LIMITED, PUBLISHERS
7/30 A, Daryaganj, New Delhi-110002, (INDIA)
Tel.: (011) 23253771, 23253472, **Telefax:** 23267437, 43551305
E-mail: contactus@newagepublishers.com • Visit us at www.newagepublishers.com
- **London** NEW AGE INTERNATIONAL (UK) LTD.
27 Old Gloucester Street, London, WC1N 3AX, UK
E-mail: info@newacademicscience.co.uk • Visit us at www.newacademicscience.co.uk
- **Nairobi** NEW AGE GOLDEN (EAST AFRICA) LTD.
Ground Floor, Westlands Arcade, Chiromo Road (Next to Naivas Supermarket)
Westlands, Nairobi, KENYA, **Tel.:** 00-254-713848772, 00-254-725700286
E-mail: kenya@newagepublishers.com

BRANCHES

- **Bangalore** 37/10, 8th Cross (Near Hanuman Temple), Azad Nagar, Chamarajpet, Bangalore- 560 018
Tel.: (080) 26756823, **Telefax:** 26756820, **E-mail: bangalore@newagepublishers.com**
- **Chennai** 26, Damodaran Street, T. Nagar, Chennai-600 017, **Tel.:** (044) 24353401
Telefax: 24351463, **E-mail: chennai@newagepublishers.com**
- **Cochin** CC-39/1016, Carrier Station Road, Ernakulam South, Cochin-682 016
Tel.: (0484) 2377303, **Telefax:** 4051304, **E-mail: cochin@newagepublishers.com**
- **Guwahati** Hemsen Complex, Mohd. Shah Road, Paltan Bazar, Near Starline Hotel
Guwahati-781 008, **Tel.:** (0361) 2513881, **Telefax:** 2543669
E-mail: guwahati@newagepublishers.com
- **Hyderabad** 105, 1st Floor, Madhiray Kaveri Tower, 3-2-19, Azam Jahi Road, Near Kumar Theater
Nimboliadda Kachiguda, Hyderabad-500 027, **Tel.:** (040) 24652456, **Telefax:** 24652457
E-mail: hyderabad@newagepublishers.com
- **Kolkata** RDB Chambers (Formerly Lotus Cinema) 106A, 1st Floor, S N Banerjee Road
Kolkata-700 014, **Tel.:** (033) 22273773, **Telefax:** 22275247
E-mail: kolkata@newagepublishers.com
- **Lucknow** 16-A, Jopling Road, Lucknow-226 001, **Tel.:** (0522) 2209578, 4045297, **Telefax:** 2204098
E-mail: lucknow@newagepublishers.com
- **Mumbai** 142C, Victor House, Ground Floor, N.M. Joshi Marg, Lower Parel, Mumbai-400 013
Tel.: (022) 24927869, **Telefax:** 24915415, **E-mail: mumbai@newagepublishers.com**
- **New Delhi** 22, Golden House, Daryaganj, New Delhi-110 002, **Tel.:** (011) 23262368, 23262370
Telefax: 43551305, **E-mail: sales@newagepublishers.com**

ISBN: 978-81-224-1464-6
C-17-02-10144

Printed in India at Kanishik Printing Press, Delhi.

NEW AGE INTERNATIONAL (P) LIMITED, PUBLISHERS
7/30 A, Daryaganj, New Delhi-110002
Visit us at **www.newagepublishers.com**
(CIN: U74899DL1966PTC004618)

Preface

The most simplified and easiest art of learning how to perform an experiment successfully is to keenly watch someone else either performing it or demonstrating it and then using the same equipment or assembly, try and repeat the whole procedure. If one does not succeed in the first attempt, then he should watch again and repeat the process; and being shown precisely the various steps involved in the experiment and pointing out what has gone wrong can rectify further failure. For majority of experiments this is one of the best methods of learning. However, it is often impracticable to follow this procedure because the specific methods are not readily available in the laboratories where one is engaged in performing the experiments.

The other method of learning is to go through a series of instructions and attempt to follow them out. This is somewhat difficult because in majority of cases it is realized that the writer often infers practical experience without keeping in mind the experience/problems of the reader. Furthermore, there are minor practical details, which are not immediately recognized by the reader or the writer. Nevertheless, this is most prevalent method of learning for conducting an experiment in a successful way, the work published in the scientific journals and periodicals furnishes the necessary information and stimulus.

However, it is often difficult to follow the experimental procedure from the published account of the scientific paper. This is frequently a weak point of the editorial system, where significantly less importance is attached to the **materials and methods** as compared to **results and conclusions**, and most authors either as a customary method or traditional way are persuaded to present this section in a very precise and condensed form.

The present volume is the first of its kind to supply sufficient practical information with minute details to enable the reader to follow and carry out the experiment for himself. This information is presented in detail, though possibly these may be too much detail for some persons and not enough for others. Mostly, those experiments that could be performed in stipulated routine class time have been incorporated in the work. However, it was felt that there were also many experiments that would take longer time, but which could not be divided into periods of smaller duration and so fit in with rather more liberal practical programme. The outstanding salient features of this work is that the procedural details of all the experiments have been described in the most lucid and readily understandable way covering the minute details of the bottle necks of each experiment.

We welcome pertinent criticism and suggestions from readers for incorporation in future contributions. It is hoped that the quality and quantity of practical information presented in this work will help to spread skill and experience from one corner to another or one lab to another.

S.C. Nigam
&
Omkar

Acknowledgements

The authors are greatly thankful to the physiologists all over the globe, whose contributions in the field of Animal Physiology have encouraged and enabled us to undertake this work in the form of a consolidated account.

At the same time, we also acknowledge the untiring efforts made by the budding scientists, Dr. (Ms.) Shefali Srivastava, Dr. Ahmad Pervez and Ms. Geetanjali Mishra without whose help it would not have been possible to complete this work in a record time.

We are also thankful to our life partners, Mrs. Sushma Nigam and Mrs. Kusum Upadhyay, respectively for ignoring our absence for greater periods of the day and taking every care of us in all the desired aspects.

S.C. NIGAM

&

OMKAR

Contents

1

Hydrogen Ion Concentration (pH)

HYDROGEN—ION CONCENTRATION

Acidity and alkalinity, often referred to as the reaction of a solution, are extremely important factors in the proper functioning of living organisms. If the reaction of our blood changes only slightly, acidosis or alkalosis will result, ending in death if allowed to go too far. In order to understand modern biology, it is quite necessary for the student to know, what is meant by hydrogen-ion concentration, and to do this; he should be familiar with some of the theory back of it.

QUANTITY FACTOR OF ACIDITY

An acid may be defined as *any substance that gives hydrogen ions when in solution.* A normal solution of an acid is a solution, which contains 1gram atom, that is 1.008grams of acid hydrogen per litre. A base may be defined *as a substance, which will give hydroxyl ions when in solution.* A normal solution of a base contains 1gram radical of ionizable OH, that is 17.008 grams per litre.

To make a normal solution of HCl, 1gram molecule of HCl, or 36.458 grams per litre of solution is needed in order to supply 1.00grams of H^+. Similarly to make a normal solution of NaOH, 40.008 grams of NaOH is required to obtain 17.008 grams of OH^- ions per litre. In like manner to make a normal solution of H_2SO_4, ½ gram molecule of this acid must be dissolved in one litre, since in

1gram molecule of H_2SO_4, there are 2gram atoms of H^+. The same principle will apply for bases, such as $Ba(OH)_2$, where there are 2 gram radicals of OH^- per gram molecule.

In usual course normal solutions are seldom used. Only standardized solutions, as designated by their normality factors, are employed. *A normality factor is actually a number, if any number of millilitres of a solution is multiplied by this factor, then it is converted into millilitres of normal solution.* For example, if 100ml of solution, whose normality factor is 0.1252 is used, only a very simple calculation is required to arrive at the conclusion that is equivalent to 100 x 0.1252 = 12.52ml of normal solution. For simplicity, the letter N is normally used for normal, and the above solution may be designated 0.1252N. A tenth normal solution is often written as 0.1N or N/10. Any solution, the exact concentration of which is known, is called as standard solution.

Leaving aside many other considerations, it is well established that, if the concentration of H^+ is known, the concentration of OH^- is also known. It thus becomes possible to express degrees of alkalinity in terms of H^+ concentration. If the negative exponent of 10 is 7, the solution is neutral; if less than 7, acid; and if more than 7, alkaline.

Since in expressing H^+ concentration, the expression 10^- is always used, Sorensen suggested leaving 10^- out and using only the numerical value of the negative exponent. He called this the pH value. The expression pH, p means power (negative exponent) and H, the Hydrogen ion concentration. Mathematically, pH is defined as *the logarithm of the reciprocal of the hydrogen ion concentration.* If a solution had a H^+ concentration of 10^{-6}, Sorensen would have said that the pH was 6. At the present time, the Sorensen system is ordinarily used, and it is customary to speek of pH values rather than H^+ concentration. The Sorensen pH scale runs from 0-14 and covers the range from normal acid to normal base. A solution with a pH less than 7 is acid and with more than 7 is basic. By means of the pH scale any degree of acidity or alkalinity found in biological materials can be expressed. It might be concluded that only whole numbers are used in expressing H^+ concentration as pH.

DETERMINATION OF HYDROGEN ION CONCENTRATION (pH VALUES)

At present, several methods are available for determining the pH of a solution. These are differentiated into two main categories, namely 1. Electrometric methods, and 2. Colorimetric method. Comparing the two, the electrometric methods are more accurate but are expansive than the colorimetric methods.

Electrometric Methods

The electrometric methods include (a) Hydrogen electrode method; (b) Quinhydrone – electrode method; (c) Glass electrode method. Out of these, the hydrogen electrode method is most easily understood and covers the entire pH range. But in comparison to the hydrogen electrode method, the one which is being used a great deal at the present time is the Glass Electrode Method.

An ordinary glass electrode consists of a thin bulb of glass filled with a standard solution of an electrolyte, such as 0.1N HCl. When this electrode is placed in the solution to be tested, a difference of potential develops between the solution inside the electrodes and that on the outside. The value of this difference depends upon the hydrogen ion concentration of the solution outside the electrode, since the inside of the electrode remains constant. The potential of the glass electrodes is determined by measuring the difference of potential between it and a calomel electrode. A calomel electrode is made up of mercury, calomel and a potassium chloride solution. KCl solution is of definite concentration; 0.1N, 1N, 3.5N and saturated solution. Such electrodes have a very definite and constant potential for a given concentration of KCl solution. Tables are available, which give pH values corresponding to differences of potential between the hydrogen electrode/ glass electrode and the four types of calomel electrodes discussed above. Because of high resistance of the glass membrane separating the solutions inside and outside the glass electrode, vacuum tube amplifiers are introduced into the potentiometer circuit to permit taking a reading. Instruments are available in the market, which read pH values directly. In the ensuring experimental exercises one such experiment, which is in use, is being described.

EXPERIMENT TO DETERMINE THE pH OF BIOLOGICAL FLUIDS GLASS ELECTRODE METHOD (CAMBRIDGE pH METER)

The biological fluids include; (1) haemolymph and gut contents of cockroach (2) the coelomic fluid of earthworm (3) the blood of *Pila* or any other gastropod (4) urine (5) milk (6) raw egg, *etc.*

Details of pH meter and its working

This apparatus (Figure 1.1) is equally versatile for determining the pH values of both the colourless and coloured fluids. It can measure from 0-14 and has an accuracy of 0.05 pH. The apparatus is housed in metallic cabinet. It consists of two input terminals. One is in the form of a socket for glass electrode, while the other one is of screw type terminal for reference electrode. The apparatus on its front panel has the following control systems.

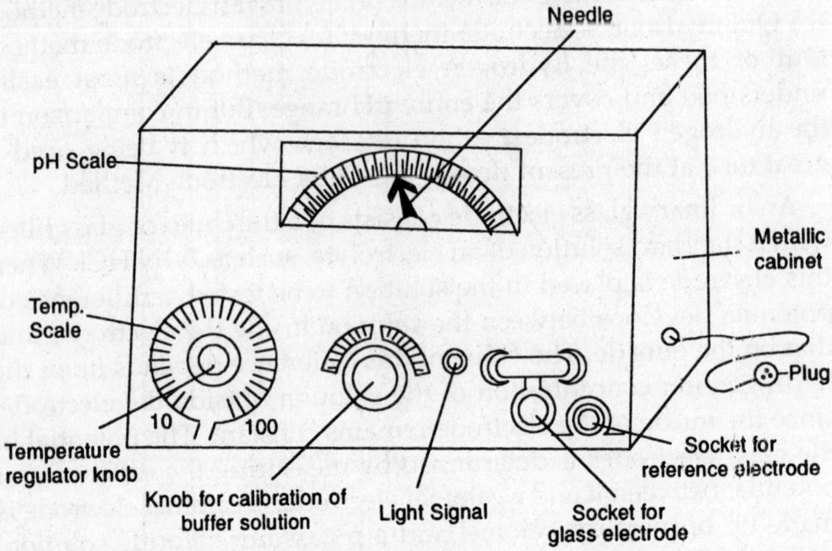

Fig. 1.1 pH Meter

1. Buffer control for standardization.
2. Temperature control compensation over a range of 10^0 - 100^0C.
3. The check switch.

The apparatus remains connected with two electrodes (Fig. 1.2) namely:

A. Glass electrode: It is made up of special glass. Its tip is very delicate, therefore extreme care is required in its handling. Its tip should not be allowed to touch rough objects, nor it should be wiped by hand or rough material. When the electrode is not in use, it should be kept immersed in a beaker containing distilled water. Glass electrode is activated by dipping its tip in 1N KCl solution for 12-24hours.

B. Reference electrode: Its bottom remains filled with 2/5 or 2/3 of its depth with crystals of potassium chloride, (KCl), and its solution. This electrode is kept moist either in KCl or distilled water. Both the electrodes remain mounted on a rubber holder on the electrode stand.

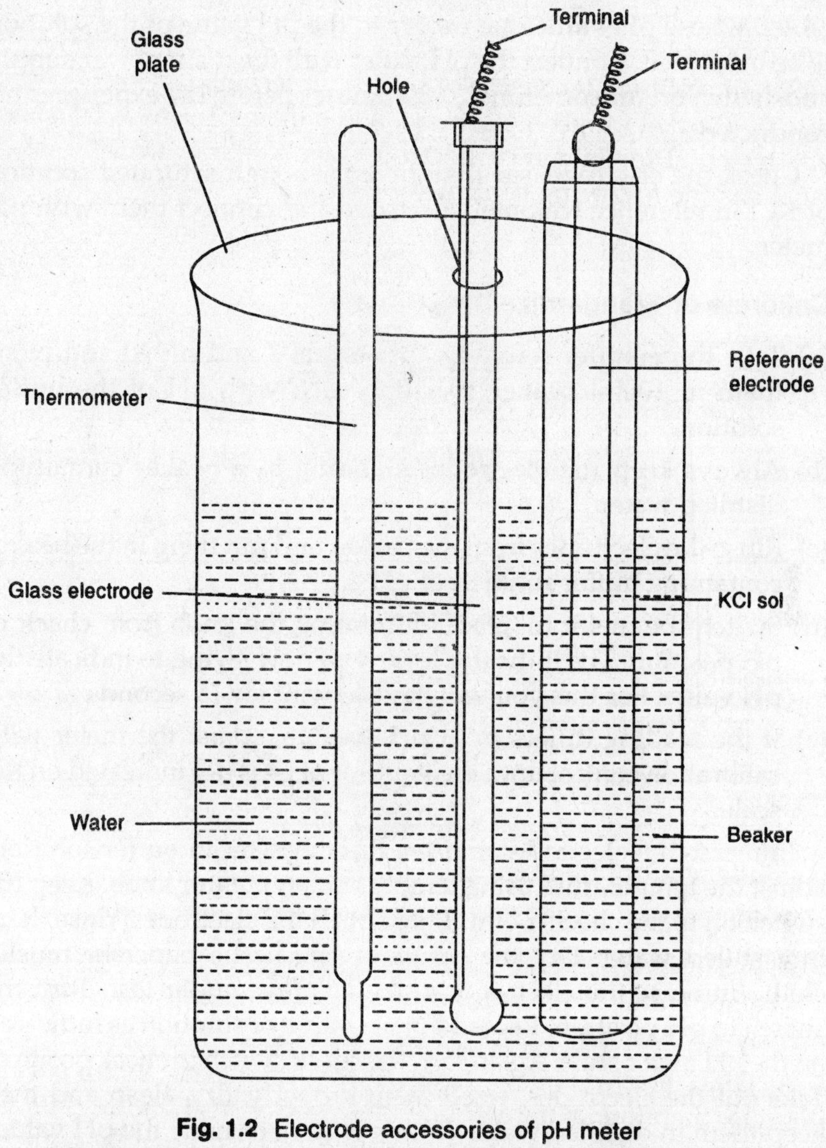

Fig. 1.2 Electrode accessories of pH meter

Operational details: Plug the terminals of the glass and the reference electrodes into the sockets concerned (Fig. 1.1). Connect the pH meter instrument with the mains and switch it on. Keep the instrument in check position all the time, whenever the electrodes are not connected or immersed in solution.

Standardisation of pH meter

Prepare standard buffer solution by dissolving the commercially available buffer tablets usually of pH 4, 7 and 9. Select the buffer tablet, whose pH values are nearer to the pH value of the solution to be measured. Connect the pH meter with the main power supply and switch on for sometime 20-30minutes before the experiment is conducted.

Check the electrode, see that there is enough saturated solution of KCl in reference (calomel) electrode and connect them with pH meter.

Calibrate or Standardize the pH meter

(a) Take three buffer solutions (pH 4, pH 7 and pH 9) and place them in 100ml beaker, marking each with pH of the buffer solution.

(b) Always keep the electrodes standing in a beaker containing distilled water.

(c) Rinse the electrodes from the beaker and put them in the beaker containing buffer solution of pH 4.

(d) Switch the meter to 'read' by rotating the knob from check to pH position. The indicator lever will now swing to indicate the pH value. For this you will have to wait for 15 seconds.

(e) If the reading differs from pH 4 value, adjust the meter with calibration control until the value of pH 4 is not indicated on the scale.

Immerse the electrodes and the thermometer in buffer solution, adjust the temperature using temperature regulator knob. Keep the switch on to the check point. Take out all the electrodes, rinse them in distilled water. Dry the tip gently by tissue paper or muslin cloth. Immerse the electrodes in the solution under test. Turn the switch to pH position. Record the pH value of solution as indicated on the pH meter scale. Again turn the check switch to check position. Take out the electrodes, wash them properly, dry, clean and then keep them in distilled water. In this way, determine the pH values of a series of solutions or biological fluid in a quick succession.

Precautions

1. Do not leave the electrodes for more than one minute in solutions that are either very acidic or basic.
2. Calibrate the pH meter with standard buffer solution near the pH value of the unknown solution.
3. Always keep the instrument in check position.

Collection of biological fluids of different animals

1. ***Blood of Pila:*** Take a living *Pila,* wash its shell well with distilled water and dry it with a clean dry cloth. Now, break the apex of the shell and invert it in a funnel standing in a test tube kept in a test tube stand. Collect enough of blood from a few *Pila.* Have a rough idea of pH value by means of litmus paper and then measure the pH with pH meter and compare the two values.

2. ***Haemolymph of cockroach:*** Take a few freshly anesthetized specimens of cockroach and wash them with distilled water, dry with filter paper. Now, hold the animal and cut its femur transversely. The haemolymph will start oozing out from the wound. Collect enough of it. Test its pH with litmus paper and then with pH meter if enough of fluid is available. Compare the two values.

3. ***Coelomic fluid of earthworm:*** Collect the coelomic fluid from a number of freshly narcotized or living specimens of earthworm as above and test the pH both by means of litmus paper and pH meter.

Note: The pH of a number of fluids including the gut contents from different regions or parts can be readily worked out by using litmus paper even when the fluid is available in very small quantity. Only the colourless fluid should be tested for pH value by litmus paper.

4. ***Gut contents of cockroach:*** Take anesthetized cockroach and dissect out the alimentary canal. Clean the adhering tissue and isolate the three distinct regions/divisions of alimentary canal, fore-gut, mid-gut and hind-gut and salivary glands. Macerate the tissues and using the wide and narrow range pH papers, find out the pH and record them in the tabulated form.

Table 1.1: Sample set for recording observations

S.No.	Name of animal	Name of fluid	Colour	pH wide range	pH normal range	Result
1.	Earthworm	Coelomic fluid	Colourless	9.0	9.3	Basic
2.						
3.						
4.						
5.						
6.						
7.						
8.						
9.						

COLORIMETRIC METHOD

This method is based on the principle that certain dyes change their original colour at very definite pH values. Such dyes are often referred to as indicators, because when they are added to a solution, the change in their colour simply indicate whether it is acidic or basic; and since they change colour when the reaction changes, they are generally used in acid and base titrations to tell when an acid or a base is neutralized. There are various indicators, which change their colour at different pH values, *e.g.* methyl orange changes its colour at a pH of 2.9 to 4.0, Congo red between 3.0 to

Table 1.2 : Showing the common and chemical names of indicators, their colour at acidic and basic ranges, pH range and preparations for measurements.

Indicator	Colour acid	Base	pH range	Preparation for use
Thymol blue (Thymo-sulfonaphthalein)	Red	Yellow	1.2-2.8 (acid range)	0.1% water solution of sodium salt
Methyl orange (Dimethyl-aminoazobenzene sulfonate)	Red	Yellow	2.9-4.0	0.1% solution in 50% alcohol
Congo red (Sodium tetrazo diphenyl naphthionate)	Blue	Red	3.0-5.0	0.5% alcoholic solution
Methyl red (o-Carboxy-benzene azodimethyl aniline)	Red	Yellow	4.4-6.0	0.05% alcoholic solution
Alizarin red (Sodium alazarin sulfonate)	Yellow	Purple	5.5-6.8	1.0% solution of sodium salt
Bromothymc! blue (Dibromo thymol sulfonaphthalein)	Yellow	Blue	6.0-7.6	0.1% water solution of sodium salt
Litmus	Red	Blue	About 7.0	Paper
Cresol red (o-Cresol sulfonaphthalein)	Yellow	Red	7.2-8.8	0.1% solution in 50% alcohol
Phenophthalein (Dihydroxy phthatlophenone)	Colourless	Red	8.3-10.0	1% solution in 50% alcohol
Cresolphthalein (o-Cresol phthalein)	Colourless	Red	8.0-10.0	0.5% alcoholic solution

5.0, sodium alizarin sulfonate between 5.5 to 6.8, litmus at about pH 7, phenophthalein between 8.3 to 10.0, chlorophenol red at pH 4.6 , phenol red at pH 6.8 (yellow) to pH 8.4 (red) (Table 1.2).

Table 1.3: Showing pH values of biological fluids as determined by the use of litmus paper.

Name	pH value
Milk	7.3
Egg white	9.4
Coelomic fluid of earthworm	8.5
Haemolymph of cockroach	8.0
Blood of Pila	8.0
Human saliva	7.0-8.3
Human urine	6.0
Fore-gut contents of cockroach	6.1
Mid-gut contents of cockroach	8.2
Hind-gut contents of cockroach	8.8
Salivary gland of cockroach	7.3

The changes in the dye colour due to change in the pH values are best demonstrated in the food vacuoles of Paramecia fed on Congo red and yeast. Similarly, if a drop of phenophthalein solution is added to some saliva in a test tube and the solution remains colourless that means the pH of saliva is less than 8.3. In another test tube take saliva and test it with litmus. If the colour is blue, that means pH is greater than 7. In other words, by this simple experiment it can be concluded that the pH of saliva is between 7 and 8.3.

For ready reference, the pH values of different biological fluids are given in Table 1.3.

2

Physiology of Metabolites

INTRODUCTION

From the biochemical point of view, four of the most important organic compounds are carbohydrates, proteins, lipids and nucleic acids. The protoplasm of the animal cell contains about 1% carbohydrates, 10%-20% proteins, 2%-3% lipids, 75%-85% water and about 1% organic compounds, such as vitamins and minerals. Carbohydrates constitute the greater percentage of food stuffs. Lipids occur in a variety of forms mainly to form membranes and to give these membranes a specific permeability. Carbohydrates and lipids together act as biological fuels (reserves of chemical energy for varied metabolic activities).

The proteins occupy a pivital position both in the architecture and functioning of the living cell. They are very closely associated with all the metabolic activities responsible for sustaining the life. The nucleic acids are DNA and RNA, which function together to manufacture the structural proteins of the cell. Proteins and nucleic acids are interrelated in their functions and both of them are essential for life.

In this chapter, the efforts have been made to detect and determine the presence of main chemical components or metabolites, such as carbohydrates, proteins, lipids and nucleic acids along with the chief classes of their constituents by means of most standard and classical biochemical and chemical tests. The tests in each section on carbohydrates, proteins, lipids and nucleic acids are preceded with a brief introduction of each of the main constituent.

It is advisable that the use of water always means distilled water unless otherwise categorically stated to use tap water.

A. CARBOHYDRATES

Carbohydrates are composed of carbon, hydrogen and oxygen. The carbohydrates will be considered first because from the chemical standpoint, they are the simplest and because they form a part of some of the more complex lipids and proteins. The carbohydrates are the most abundant constituents of many food items and are the source of large part of the heat and mechanical energy of the body. Although the carbohydrates are usually thought as a source of heat and energy, it should not be forgotten that in animals, these carbohydrates occur as part of the protoplasm, where they are the richest source of energy required for general metabolic activities of living organisms. These carbohydrates are identified by a variety of methods as described below:

CHEMICAL IDENTIFICATION BASED ON COLOUR REACTIONS

A number of specific colour tests are available, as detailed below:

1. Molisch's test (α-naphthol reaction)

Take 4-5ml of 4% glucose solution in a test tube and add 2-3 drops of Molisch's reagent to it. Mix the two by shaking the test tube and then pour very carefully from the side, a few drops of concentrated H_2SO_4, so that a layer of acid is formed at the bottom of the test tube. Wait and watch the development of a violet or reddish-violet coloured ring, indicating the presence of sugar.

This test is based on the formation of furfuraldehyde as a result of the action of acid on carbohydrate and on its reaction with α-naphthol, resulting in the development of a characteristic colour.

Preparation of Molisch's reagent

This is prepared by dissolving 5grams of α-naphthol in 100 ml of 95% ethyl alcohol or absolute alcohol. In the absence of α-naphthol, 3-4 drops of 5% alcoholic solution of thymol can be used.

2. Moore's test (It is based on the action of alkali with carbohydrates).

Take 3-4ml of 4% glucose solution in a test tube and add a few drops of saturated solution of NaOH or KOH to it. Place the test tube in a boiling water bath so as to bring it to boiling point. The development of a brown colour and an odour of carmel is an indication for the presence of carbohydrate. During baking, if too much soda is used as the leavening agent, a tan product will be formed by the action of alkali on the sugar present.

3. Benedict's test

Take 4-5ml of Benedict's reagent in a test tube and add 8-10 drops of 1 or 2% sugar solution to it. Heat the mixture by placing the test tube in a boiling water bath for 3-4minutes and thereafter allow the test tube to stand and cool down. If a precipitate of red, reddish-brown, yellow or green colour is developed, that means the sugar is present.

It is an excellent test for detecting the presence of reducing sugars, such as glucose and fructose, as these possess an aldehyde or ketone group, which reduces the ions of metals, such as bismuth, copper, iron and silver. The underlying principle is based on the quality of reducing sugars to convert a blue cupric hydroxide solution on being heated into in an alkaline medium into a yellow or orange-red or red precipitate of cuprous oxide.

Preparation of Benedict's reagent

Take 800ml of distilled water in a large flask and add 86.5 grams of sodium citrate and 50 grams of sodium carbonate to it. Heat the mixture till both the compounds are not dissolved completely. Make the volume of this mixture upto 425ml by adding more of distilled water. In another flask, dissolve 8.5 grams of copper sulphate in 50ml of distilled water, shake well to make a homogenous solution. Take a large beaker and transfer these two solutions by constant stirring and finally make up the volume of the whole mixture to 500ml by adding more of water.

4. Fehling's test

Take about 1ml of Fehling's reagent in a test tube and add to it only one or two drops of sugar solution under test at time. After each addition, the mixture should be heated. The development of a characteristic brick red or reddish precipitate should be regarded as an indication for the presence of reducing sugars. Fehling's solution should always be tested because sometimes this solution, on prolonged storage, gives a similar test on simply boiling with a few drops of water.

Preparation of Fehling's reagent

It is prepared shortly before use by mixing the followng two solutions in equal volumes.

(i) **Copper sulphate solution:** In a 250ml flask, dissolve 17.5 grams of copper sulphate in some distilled water and make the volume upto 250ml by adding more of distilled water.

(ii) **Alkaline tartarate solution:** Take another flask of same capacity with some distilled water and dissolve 62.5 grams of potassium hydroxide (KOH) and 86.5 grams of Rochelle salt in it, mix the two and make the volume upto 250ml.

5. Barfoed's test

Take about 1-2ml of 3-4% solution of sugar (glucose, fructose or galactose) in a test tube and add about 3ml of freshly prepared Barfoed's reagent to it. Place the test tube in a boiling water bath for only a few minutes. The development of a red precipitate will indicate the presence of sugars. Prolonged boiling should be avoided. It is a specific test for the monosaccharides but not a proper test for detecting the presence of sugars in the urine samples, as it works only in alkaline medium.

Preparation of Barfoed's reagent (Tauber-Kleiner's Modification)

Dissolve 24 grams of copper acetate in 450ml of boiling water, without care, even if a precipitate is formed. Add 25ml of 8.5% lactic acid to it and shake it thoroughly well till all the precipitate is dissolved completely. Allow it to cool and make the volume upto 500ml by addition of more of distilled water. The reagent is now ready for use. It should be stored in a cool place in a glass stoppered bottle.

6. Potassium permanganate test

Take 4-5ml of saturated solution of $KMnO_4$ in distilled water in a test tube, add a concentrated solution of sugar to it and heat it. The disappearance of the red colour will indicate the presence of sugar. It is the most simple test and can be used as a preliminary test.

7. Nylander's test

Take 4-5ml of 3-4% sugar solution under test in a test tube and add a small quantity of Nylander's reagent to it. Heat it to a boiling point by placing the test tube in a boiling water bath. If a black precipitate of metallic bismuth is produced, that means the reducing sugars are present.

Preparation of Nylander's reagent

This is prepared by dissolving 4grams of bismuth subnitrate and 8grams of Rochelle salt in 200ml of 10% solution of KOH and heating the mixture. Allow it to cool and then filter and store for further use.

Note: Nylander's reagent has certain added advantages over other copper containing reagents meant for testing the presence of sugars in urine samples, as it is not reduced by uric acid and other urine constituents, which usually reduce copper.

8. Cole's test

Take 5ml of 5% glucose solution and add just sufficient quantity of anhydrous sodium bicarbonate to it. Add 3-4 drops of glycerine and 3-4 drops of 5% copper sulphate solution to this. Heat the mixture over a boiling water bath for a few minutes. If a reddish yellow or orange coloured precipitate is developed that means the reducing sugars are present.

9. Seliwanoff test

This test is meant for differentiating aldoses from ketoses and D-fructose. Take just 0.5ml of sugar solution (fructose or ketose sugar) in a test tube and add 3ml of Seliwanoff's reagent to it and boil the mixture over a boiling water bath for a few minutes only. It will give a red colour with ketoses but not with an aldose sugar.

Preparation of Seliwanoff's reagent

This reagent is prepared by just dissolving 0.5g of resorcinol in 100ml of dilute hydrochloric acid. HCl should be diluted by mixing 1 vol of HCl with 2 volumes of water.

10. Picric acid test

Take 5ml of 2-3% of glucose solution in a test tube and add 2-3ml of saturated aqueous solution of picric acid to it and approximately

1ml of 10% solution of sodium carbonate. Warm it gently over a low flame. If a mahogany red colour is developed, then it should be regarded as an indication for the presence of reducing sugars (glucose). This colour is developed due to reduction of picric acid into picramic acid.

11. Iodine test for starch

Take 3ml of 1% starch solution in a test tube and add a few drops of Lugol's iodine to it. If a blue colour is developed then it should be taken as an indication for the presence of starch. This can be further confirmed by heating the mixture, the colour will disappear but will reappear on cooling.

Preparation of Lugol's iodine

Take an amber coloured glass stoppered bottle and dissolve in it 1gram of iodine and 2grams of potassium iodide in 100ml of distilled water.

12. Benedict's test for starch

Take about 20ml of 1% starch solution in a 100ml beaker and add 10-12 drops of concentrated HCl to it and bring the contents to boiling point by placing the beaker in a boiling water bath for about 3-5 minutes. By this treatment, the starch is hydrolyzed into its sugar molecules. The mixture is allowed to cool and the end products of hydrolysis are neutralized with a solution of sodium carbonate.

Now, take a few ml of this hydrolysate in a test tube and add 3ml of Benedict's solution to it. Heat the mixture to boiling point in a water bath for 3-4 minutes and then allow the mixture to cool down. If a reddish brown precipitate appears, then it should be taken as an indirect evidence for the presence of starch.

13. Iodine test for cellulose

Take a few cotton threads in a test tube and dissolve either in Schweitzer's reagent or in Gross and Bevan's reagent and add a few drops of concentrated H_2SO_4 to it. Now, add a few drops of Lugol's iodine. If a blue colour is developed that means cellulose is present. This test is based on the conversion of cellulose into amyloid, which gives a blue colour with Lugol's iodine. Prolonged treatment with concentrated H_2SO_4 should be avoided, else the amyloid formed will be converted further into glucose.

Preparation of Schweitzer's reagent: It is an ammonical solution of copper hydroxide ($Cu(OH)_2$).

Preparation of Gross and Bevan's reagent: It is a solution of zinc chloride ($ZnCl_2$) in HCl.

14. Campbell's chitoson test for chitin

Chitin comprises the bulk of the exoskeleton of crustaceans, insects, *etc.* Dissect out a portion of cuticle of either a mature adult insect or the larva and clean it with cotton swab to remove any adhering tissue. Place it in a test tube and treat it with saturated solution of KOH at 55°C over a liquid paraffin bath for about two hours. Remove the test tube and allow it to cool down to room temperature. By this treatment the chitin, which is a glucosamine polysaccharide is converted into chitoson. Remove the treated piece of cuticle and wash it thoroughly with distilled water. Place it in a cavity block or watch glass and treat with 4-5ml of 3% acetic acid at room temperature. The chitin part (chitoson) will undergo complete dissolution indicating its presence. This is further confirmed by adding a few drops of 1% sulphuric acid in the same watch glass or cavity block resulting in the formation of a white precipitate, which is readily seen against a black paper background. If the material undertest is not chitin, it will dissolve in KOH. If on counter treatment with acetic acid anything is left, that also indicates its non-chitinous nature.

B. LIPIDS

The lipids are important constituents of protoplasm. They are characterised by a low solubility in water but highly soluble in organic solvents, such as, ether, acetone, chloroform and other fat solvents. In the animal body, they form the main store of reserve food supply, being derived from fat and carbohydrate diet. The higher contents of certain lipids in the brain and nervous tissue indicate the importance of these compounds to life.

PHYSICAL AND CHEMICAL IDENTIFICATION OF LIPIDS

1. Solubility test

Take 5ml of water, dilute acetic acid, dilute KOH, ethanol, benzene, solvent ether, acetone, chloroform and carbon tetrachloride in seven test tubes separately. Add a few drops of cottonseed oil or olive oil or any other natural food oil in each test tube. Shake well, allow the tube to stand and observe that the mixture becomes homogeneous.

2. Emulsification test

Take 5ml of distilled water, add a few drops of the emulsifier (bile salt) and the same amount of oil in a test tube. Shake well to mix the contents. The lipid is dispersed forming an emulsion in the form of micelles, *i.e.* fats, which are broken up into minute droplets in water. This is brought about as a result of the lowering of the surface tension between fat and water. The bile salts or other substances that bring about emulsification are referred to as emulsifiers.

3. Paper spot test

Rub a bit of fat sample, say olive oil or mustard oil, on a piece of paper. If the spot on the paper assumes a translucent appearance, that means the sample under test is a fat or oil, *i.e.* a lipid.

4. Saponification test

Take 100ml of distilled water, 50ml of 10% ethanolic solution of NaOH or KOH and add about 10grams of fat or oil in a 250ml flask or beaker. Reflux this over water bath under a reflux condenser for an hour or so, till the saponification is complete. Placing a few drops of the saponified mixture on the surface of water can ensure this. If oily droplets do not separate, then the reaction is complete. The mixture consists of glycerol, NaOH, water and sodium soap. Now proceed further with this soap solution as follows:

(i) Add powdered sodium chloride to the point of saturation with constant stirring to 20ml of soap solution in a beaker. The soap is separated from its sodium component and being insoluble rises to the surface.

(ii) Take a few ml of the saponified mixture in a test tube and add a few drops of calcium chloride solution (10%) to it and mix well. If a white precipitate that dose not dissolve readily on dilution is formed, that means the lipids are present. The precipitate corresponds to the soap of calcium, which is hard. This can also be performed with 10% solution of magnesium sulphate, where the insoluble soap of magnesium is formed. Calcium soaps are formed with hard water where the hardness is due to the calcium ions.

5. Test for unsaturated lipids

Take a few ml of ethanol in a test tube and add a few drops of olive oil or any other edible oil to it. To this add 0.4% bromine water drop by drop till an orange colour is developed. This indicates the presence of double bond. The intensity of the colour developed corresponds to the degree of unsaturation. If the same test is performed with acetic acid; no colour will be developed, as it is a saturated acid.

6. Lieberman-Burchard test for cholesterol

Take 2ml of chloroform in a test tube and dissolve a pinch of any sterol, e.g. cholesterol, in it. To this add at first 8-10 drops of acetic anhydride and then a drop or two of concentrated H_2SO_4 from the side. A greenish or blue colour will be developed indicating the presence of cholesterol.

7. Odour test

Place 0.5 to 1.0 gram of finely divided potassium bisulphate ($KHSO_4$) in an air-dried test tube and add 3-4 drops of olive oil to it. Heat it gently. The heating should be carried out progressively. A characteristic odour of acrolein will be felt, which is on account of the dehydration of glycerol moiety resulting in the production of acrylic aldehyde or acrolein.

8. Sudan black B test

Take a few ml of distilled water in a test tube and add a drop or two of the olive oil to it. Shake the tube so that the mixture is converted into an emulsion. Place a small drop of this emulsion on a microslide and add to it a little bit of Sudan Black B stain, mix and put a cover glass. Examine under microscope and note that only oily droplets are stained, this happens so because the stain is very specific for lipids.

9. Sudan III test

Take a few ml of either olive oil or its emulsion in a test tube and add a few drops of Sudan III solution (in ethanol) to it, a red colour will be developed indicating the presence of lipids.

C. PROTEINS AND AMINO ACIDS

Protoplasm is generally regarded as being composed primarily of proteins. It is for this reason, they are considered to be the most important compounds in respect to chemistry of life. For animals the proteins are extremely important as compared to other metabolites, like the lipids and carbohydrates, because without proteins, animals are unable to form new protoplasm. Since it is not possible for an animal to live on carbohydrates or lipids or both without proteins, so the proteins seems to have a unique place and functions in the economy of animal life.

Proteins are extremely complicated compounds. They are made up of many simple molecules called amino acids. The amino acids are called the alphabet of protein chemistry. The proteins are differentiated into simple, conjugated and derived proteins and are identified by a number of colour reactions, whereas the amino acids are identified by paper chromatography or thin layer chromatography, which will be discussed in the respective chapters.

It should be noted that the colour reactions of proteins detailed below are specific tests not for proteins but for certain structures usually found in proteins. Therefore, ordinarily one colour test alone should not be regarded as a conclusive evidence for the presence of proteins. That means at least a few selected colour tests should be performed, which specifically respond to a characteristic group of proteins.

CHEMICAL IDENTIFICATION OF PROTEINS

1. Preliminary test for proteins

Take an air-dried test tube and place some powder of dried egg albumin in it. Now, suspend a strip of moist red litmus paper inside it and cover its mouth with a piece of filter paper moistened with a solution of lead acetate. Gently heat it over a low flame and note the changes as the reaction proceeds. The powder will be charred indicating the presence of carbon, the red litmus will be turned blue due to the ensuing fumes, indicating the presence of ammonia and the lead acetate paper will turn black due to the presence of sulphur. The appearance of the moisture inside the walls of the test tube will indicate the presence of hydrogen.

2. Biuret test

Take 2-3ml of protein solution, *e.g.* the egg albumin, casein, gelatin, peptone, in a test tube and add an equal volume of 10% NaOH solution, mix it thoroughly well and add 0.5% copper sulphate solution to it drop by drop with a constant mixing until a purplish-violet or pinkish violet colour is developed. The colour developed varies with the nature of the protein, as for instance proteoses and peptones give a definite pink, while the colour produced with the gelatin is more or less blue. Care must be taken to avoid the excessive use of copper sulphate ($CuSO_4$), else the violet or pink colour will be obscured by the blue precipitate of copper hydroxide (Cu $(OH)_2$) formed in the reaction. This test is so named because the compound biuret also gives a similar colour reaction.

3. Million's test

Take 5ml of diluted solution of such protein sample, *e.g.* egg albumin (10%), gelatin (2%), casein (2%), peptone (2%), in separate test tubes and add 4-5 drops of Million's reagent (freshly prepared) in each. In each case mix the two thoroughly well and heat the mixture gradually raising its temperature to the boiling point. The development of a red or flesh colour is an indication of the presence of protein. If a clear indication of development of colour is not apparent in the beginning then add 2-3 drops more of the reagent and heat it again. Excessive use of reagent should be avoided else it may result in the development of a yellow colour, which has to do nothing with the test. Million's reagent is a mixture of mercuric nitrate and nitrite and is made by dissolving mercury in concentrated HNO_3. Only proteins containing tyrosine give a positive test. Hydroxy-phenol group-C_6H_4OH of tyrosine is responsible for this test. The red colour is due to a mercury salt of nitrated tyrosine.

Preparation of Million's reagent

Dissolve 10gm of mercury in 20ml of concentrated HNO_3, when thoroughly dissolved, dilute with 60ml of water.

4. Cole's modification of Million's test

Take 1ml of diluted egg albumin solution in a test tube and add concentrated sulphuric acid (H_2SO_4) to it. Boil gently for half minute. A yellow precipitate will appear first. Cool it under tap water and add 1% sodium nitrite ($NaNO_2$) solution. Again warm it gently, the precipitate formed earlier or the colour of the solution will turn red. The same test with phenols also produces almost similar results.

5. Xanthoproteic test

The word ***Xantho*** stands for yellow, so the test is often referred to as the Yellow Protein Test. For this, take 4-5ml of the diluted solution of the protein, *e.g.* egg albumin, gelatin, casein, proline, peptone, tyrosine, *etc.* separately in each test tube and slowly add 1ml of concentrated HNO_3 in each. Heat it gently. A white precipitate is first formed, which turns yellow and partially undergo dissolution on further heating or boiling. Now, cool the solution by applying running tap water on the test tube and add either saturated solution of sodium hydroxide (NaOH) or ammonium hydroxide (NH_4OH) drop by drop until the solution becomes alkaline and the colour changes to a burnt orange or orange alone. This test responds positively for proteins containing amino acids with benzene ring, such as, tyrosine, phenylalanine and tryptophan. Benzene rings are apparently nitrated and the nitrated compounds produce the burnt orange colour on interaction with alkali.

6. Hopkin's-Cole test for proteins containing tryptophan

Take 4-5ml of protein solution in a test tube and slowly add a few drops of a solution of glyoxylic acid (CHO.COOH) to it and thereafter a layer of concentrated sulphuric acid (H_2SO_4) is placed under this solution, by passing the acid from the sides of the tilted test tube. This results in the development of the violet ring at the juncture of the two liquids, indicating the presence of tryptophan in the protein sample under test. This colour is developed due to the indole group of tryptophan, which is converted by the aldehyde group of the glyoxylic acid into a violet compound.

7. Liebermann's test for proteins containing tryptophan

Take a small quantity of solid protein sample in a test tube and add a few drops of concentrated HCl to it. Heat the mixture to boiling point and then add a few drops of dilute sucrose solution. If the violet colour is developed that means tryptophan is present. The aldehyde group responsible for the development of the colour is made available by the action of HCl on sucrose.

8. Ehrlich's p-Dimethylamino benzaldehyde test for proteins containing tryptophan

Take 3-4ml of diluted solution of protein in a test tube and boil it with a small quantity of concentrated HCl. Thereafter, add a few drops of p-dimethylamino benzaldehyde dissolved in 10% H_2SO_4. The development of red to violet colour indicates the presence of tryptophan. The colour changes to blue upon the addition of a few drops of 0.5% $NaNO_2$ solution.

9. Acree-Rosenheim test for proteins containing tryptophan

Take 2-3ml of protein solution in a test tube and add a few drops of very dilute formaldehyde (HCHO) solution. Thereafter, add a few drops of concentrated H_2SO_4 from the side of the test tube. Avoid shaking of the test tube. If a violet ring appears, that means tryptophan is present. The aldehyde group responsible for the production of the violet colour is made available from the formaldehyde. This is a very important test for confirming the presence of formaldehyde in the milk because milk vendors use it as a preservative so that surplus milk is not spoiled.

A more simplified method for ascertaining the presence of formaldehyde (HCHO) in the milk sample is just to heat the milk sample with a few drops of concentrated HCl. If a violet or purple colour is developed, it means the sample contains formaldehyde.

10. Ehrlich's Diazo test for proteins containing tyrosine

Take 2-3ml of protein solution in a test tube and add a few drops of Ehrlich's Diazo reagent to it. Make it alkaline by adding a few drops of NH_4OH, if a red or orange colour appears, it means tyrosine or histidine is present. The presence of former produces a light colour while the latter gives a red to orange colour.

Preparation of Ehrlich's Diazo reagent

This reagent is made by mixing 2% $NaNO_2$ solution with sulfaniline acid dissolved in 20% HCl solution.

11. Sulphur test

This test is meant to detect the presence of S-S Group (Cystine) and S-H Group (Cystiene), the two amino acids that contain sulphur.

Take a few ml of protein sample in a test tube or simply a few hair clippings and boil with 4-5ml of 50% KOH solution for 1-2 minutes. Now, add a few drops of 10% lead acetate $(CH_3COO)_2Pb)$ solution to it, if a black precipitate is formed that means sulphur containing amino acids, such as, cystine and methionine are present. The action of strong alkali results in the separation of the sulphur from the amino acids, which in turn forms potassium sulphide (K_2S). This then reacts with lead acetate ($(CH_3COO)_2Pb$) to form lead sulphide (PbS), which appears in the form of a black precipitate.

12. Molisch's test for protein containing sugars

Take a few ml of protein solution containing sugar and add a few drops of Molisch's reagent to it and thereafter, slip a little bit of concentrated H_2SO_4 from the side of the test tube. The formation of a violet ring should be considered a positive test. This is a very satisfactory test for glycoproteins. Molisch's test is primarily meant for carbohydrates. Molisch's reagent is an alcoholic solution of α-naphthol.

Preparation of Molisch's reagent

Dissolve 5gm of α-naphthol in 100ml of 95% ethyl alcohol.

13. Ninhydrin test

This is also referred to as triketohydrindene hydrate reaction. For this, take 5ml of protein solution in a test tube, which should be approximately neutral, *i.e.* in between pH 5 to pH7 and add 0.5ml of 0.1% solution of Ninhydrin (triketohydrindene hydrate) to it. This should be followed by heating to boiling point for about one to two minutes and finally allowed to cool. If a blue colour is developed that means protein is present.

14. Precipitation test

There are several methods by which the proteins are precipitated.

(i) **Precipitation by treatment with heavy metals:** Take a set of six test tubes and place 2-3ml of diluted solution of egg albumin in each test tube. Take one test tube and add mercuric chloride ($HgCl_2$) solution to it drop by drop until an excess of reagent is added. The formation of the precipitate should be regarded as an indication for the presence of protein. Similarly, test the solution in remaining test tubes with any other heavy metal, like lead acetate, mercuric nitrate, silver nitrate, ferric chloride, barium chloride, copper sulphate, zinc sulphate, *etc.* The solutions under test should be very dilute.

(ii) **Precipitation by treatment with acid or alkaloid reagents:** Prepare a set of test tubes each containing 2-3ml of diluted solution of egg albumin. Take one test tube and add drop by drop but slowly the picric acid solution to it and observe the formation of the precipitate, an indication of the presence of protein. Similarly, test the solution in remaining test tubes separately with tannic acid, ferrocyanic acid, metaphosphoric acid, sulphosalycylic acid, phosphotungstic acid, phosphomolybdic acid and potassio-mercuric iodide. While performing the test with last three reagents, the protein solution should be acidified first with HCl and tested with litmus for acid redicals.

(iii) **Precipitation by treatment with organic reagents:** Take 2-3ml of diluted solution of egg albumin in a test tube and add a few drops of 1% solution of Trichloracetic acid (TCA) and heat. The formation of a precipitate indicates the presence of protein.

(iv) **Precipitation by treatment with alcohol:** Take 4-5ml of electrolyte free solution of egg albumin in a test tube and add 9-10ml of 95% ethyl alcohol to it. Mix the two thoroughly well. Now, add either a pinch of sodium chloride (NaCl) or a few drops of sodium chloride solution to it. The formation of a precipitate is an indication of the presence of protein.

Preparation of electrolyte free solution: Take a dialyzing bag and place in it the egg albumin solution and add a drop of toluene to act as a preservative. Seal the mouth of the bag with a thread and immerse it in a large beaker containing distilled water (preferably double distilled) and allow it to stand as such for several days. During this period the water of the beaker should be changed frequently. Remove the bag take out the albumin solution in a beaker. Filter it and test for the absence of chloride before using it for the test.

(v) **Precipitation by treatment with ammonium sulphate ($(NH_4)_2SO_4$):** Take 4-5ml of egg albumin solution in a test tube, add a large quantity of ammonium sulphate ($(NH_4)_2SO_4$) salt to it and observe the formation of the precipitate. Since the proteins are hydrophilic colloids, the addition of ammonium sulphate removes the water, the large quantities of which aid in their dispersability, thus resulting in the formation of the precipitates. The process is also referred to as Salting out.

(vi) **Precipitation by treatment with heat:** Take 3-4ml of undiluted white of an egg in a test tube and place in a boiling water bath and observe the formation of a white coagulum. This heating denatures the salt free albumin and changes its nature from lyophilic to a lyophobic colloid state. The particles of the colloids are prevented from uniting due to the electric charges present on them. Coagulation and flocculation, which are often compared with denaturation are simple visible manifestations of the changes in the protein molecules brought about by denaturation.

(vii) Precipitation at isoelectric point of protein:

Theory: The proteins are colloidal substances that is why they are capable of taking up large quantities of water, which aids in their dispersability. These remain in colloidal state because of the electric charges present on the colloidal particles. The charges are either positive or negative depending upon the hydrogen-ion concentration of the solution. In neutral solution most of the proteins have a negative charge, when an acid is added to it, the negative charge is first reduced and thereafter become positive. The point of equilibrium that is when the negative and the positive charges are equal is known as the isoelectric point and it is at this point that the proteins are most readily precipitated.

Procedure: Take 4-5ml of milk in a test tube and add just sufficient quantity of HCl, so as to bring the pH of the solution to about 4.7, the isoelectric point of casein and observe its precipitation. Similarly, gelatin and many other proteins that are soluble in water even at their isoelectric points are made to precipitate more readily if some precipitating agent like that of an alcohol is added.

3

Histochemistry of Metabolites

(Localisation of chemicals in situ)

Histochemistry deals with the identification or localization of the reactive compounds or their enzymes at their normal sites in the cells or tissues by means of reagents and stains, which give specific visible colour reactions indicative of presence of the substances or enzymes. The intensity of visible colour is an indication of their effective concentration or activity.

THE FUNDAMENTAL BASIS OF HISTOCHEMISTRY

The histochemistry in a broad sense answers two questions; (i.) it gives information about the gross chemical composition of cells or tissues, and/or (2) an understanding about the degree of activity of cells or tissues under study.

Once pathologists were concerned over the diagnosis of a tumour in the neck of a patient. On examination of the tissue sections of the tumour, it appeared to be composed of either ganglion cells (*i.e.* to be diagnosed as a carotid body tumour) or of fibroblasts (*i.e.* in a fibro-sarcoma). The expert opinion was divided. The only way out was to go in for histochemical screening because qualitative histochemistry was characteristic for one or the other type of cell. One method conclusively showed the presence of high concentrations of gangliosides indicating that it should be a carotid body tumour. In the second case, a secondary growth was found in a lung. From the histology alone, it was difficult to ascertain whether the primary growth was in the liver or adrenal. The histochemical test revealed a high concentration of steroids in the cells of secondary growth, thus strongly indicating that primary tumour was found in the adrenal. In both these examples, histochemistry was applied to identify the chemical characteristic of the tissue and aided the diagnosis based on histology.

THE DEGREE OF ACTIVITY OF CELLS OR TISSUES

It all depends on the availability of the active groups, which are actually available to the dye for reaction. The major differences between the cells or tissues lie in the degree to which their active groups or enzymes are involved. This activity forms the basis of different physiological and pathological conditions occurring within a single tissue or organ. This gives precise information about the extent the reserve is being called into play at the time of biopsy.

The major steps involved in histochemistry are given below:

THE PREPARATION OF TISSUE SECTIONS

This is an important aspect, because the sections prepared in different ways even when subjected to the same staining method yield very different results, depending on how the tissue and sections have been processed. Therefore, for the preparation of sections, it is very important to ensure that the tissue is adequately hardened and autolysis completely stopped.

The different revelations using the same staining methods on sections prepared by using different features is due to the alterations in histochemical nature of tissue as there are some enzymes, which remain unaffected on being fixed in formalin, while the other on being treated into alcohol and acetone undergo denaturation, or their lipid-protein bonds are affected, which destabilize the membrane structure and the control of enzymatic activity. For the histochemical detection of such enzymes, the test is applied on fresh frozen sections. Some of the methods of fixation and preparation of tissue sections, which are routinely used in laboratories for demonstration purposes, are as under:

1. Preservation of tissue before processing for sectioning

Sometimes it is not possible to process the freshly taken out tissue immediately for preparing the sections, in such a case the tissue is stored after fixing it in suitable fixative for avoiding the postmortem changes. There are two simple ways of storing the tissue: (i.) In first case, the tissue fixed in formol-calcium can be stored in either the glycerine (A-R grade) solution prepared by mixing equal volumes of pure glycerine and distilled water. This practice helps to keep the enzymatic activity upto a period for at least 8-9 months (ii.) In the second case, the tissue fixed in calcium-formol is preserved in sucrose solution, prepared by dissolving sucrose 15.0gm, acacia 0.5gm and 50ml of distilled water.

2. Chemical fixation of tissues and its embedding in hard matrix by Gomori's Method

This is suitable for fixation of certain enzymes and involves the following steps:

(i) Dissect out the piece of tissue of approximately 2.4mm and heat it with chilled acetone (at 4°C).

(ii) Transfer in a cavity block containing a mixture of ether-alcohol of equal volumes and keep it in refrigerator for at least 2hours.

(iii) Again transfer it in another cavity block containing 10% solution of nitrocellulose and leave it in refrigerator for 3-4hours so as to ensure complete infiltration of tissue with nitrocellulose.

(iv) Replace the nitrocellulose with either benzene or chloroform and wait for ½ hour to 1hour.

(v) Now, transfer the tissue in embedding pan already containing melted paraffin kept in a temperature bath or incubator set at 52°C, (the melting point of wax) for ½hour to 1hour or more depending upon the nature of the tissue and its thickness.

(vi) Make suitable paraffin blocks of the tissue pieces, store them in refrigerator till the sections are not prepared and flattened on slides as per routine microtomy technique.

Note: The simplest way of above procedure is to fix the tissue in chilled acetone for 9-12 hours, followed by two treatments with cold petroleum -ether, each time for 60min and thereafter to bring it down to room temperature followed by embedding in paraffin wax (m.p. 45°C) for making the blocks.

3. Preparation of sections by Freeze Drying technique.

The important feature of this method is that autolytic processes are spontaneously brought to stand still by 'quenching' the small pieces of tissue approximately 1-2mm with isopentene cooled by liquefied gases, such as liquid nitrogen at a very low temperature (between -160°C to -190°C). Water is removed from the tissue at -40°C to -70°C by the process of dehydration using ethyl alcohol, methanol, acetone, *etc.* The temperature of dried tissue is brought to room temperature and embedded in paraffin. In this way the tissue is fixed in a hard matrix, so that paraffin sections can be cut with a conventional microtome. The slides are dewaxed with xylol and the sections are fixed in the dried state in absolute alcohol.

4. Preparation of sections by Freeze-Substitution technique

This technique involves the following few steps:

(i) Take small pieces about 1-3mm and quench them by pulling in a mixture of propane-isopentane brought to a temperature between -170°C to -175°C with liquid nitrogen involving vigorous stirring for half a minute.

(ii) Remove water from frozen pieces by substituting methanol or ethanol already cooled to -70°C. These blocks can be stored in dry ice (solid SO_2) chest for 1-2 weeks.

(iii) Transfer the tissue blocks in a refrigerator and wash them with at least three changes of absolute alcohol in duration of 12-15hours.

(iii) Replace absolute alcohol with chloroform and allow the blocks to remain in it for 5-10hours.

(iv) Now, replace chloroform with fresh chloroform at 0°C and wait till the room temperature is not attained.

(v) Transfer the tissue block in molten paraffin wax (m.p.52°C) in an oven for 30minutes.

(vi) Prepare paraffin block of the tissue and cut sections with a conventional microtome.

5. Preparation of sections by Freeze-Sectioning technique

In this technique, the autolysis is stopped and tissues hardened simultaneously by simply freezing the tissue pieces and cutting sections of the frozen blocks. Apart from many shortcomings, this procedure is widely recommended for all histochemical investigations. However, for certain reactions it becomes necessary to fix the sections specifically for demonstrating the presence of certain compounds.

The main steps involved in this technique are as under:

(i) Chilling of the tissue

It is carried out in a chilling bath, which is prepared by pulling small chips of solid CO_2 in alcohol in a chilling bath until a saturated solution is formed. At this stage, the temperature should be -70°C.

Thereafter, a small beaker having 30-50ml of hexane already cooled to -65°C is placed inside the bath. At this stage more of CO_2 is added to bath to maintain the temperature.

The tissue under investigation is cut into small pieces of suitable size about 5mm and these are then placed in hexane without getting an access to sides of beaker. The tissue pieces should not be allowed to remain immersed in hexane for more than 2minutes, the maximum extent. After this treatment, transfer the tissue pieces with the help of pre-cooled forceps to a dry tube at -70°C. Care must be taken that the tissue does not warm up during transference. The tube should be corked and encased in solid CO_2 in the flask for storage.

The sections of these ***chilled blocks of tissue pieces*** or fresh frozen tissue are cut with the help of cryostat microtome. It is, therefore, necessary that before using the cryostat, one must have a clear understanding of its various components and *modus operandi*.

Nowadays, cryocut sections are most commonly used in histochemistry of metabolites (carbohydrates, lipids, proteins and nucleic acids) and enzymes. The sections of both the fixed and unfixed (fresh) tissues are prepared by cryocut cryostat microtome. The sections of the fixed tissues prepared by cryostat are referred to as ***fixed frozen sections,*** while those of unfixed fresh tissues as ***fresh frozen sections.***

The use of freezing microtome is restricted to orthodox laboratories, while in most modern laboratories, the cryocut or cryostat microtomes are being used for routine work.

The Cryocut Cryostat Microtome
(various components and *modus operandi*)

The cryostat is a rotary microtome, which has been mounted in a refrigerated cabinet, which itself is provided with an insulated transparent sliding door that carries the coils of wires embedded in its matrix and meant for keeping the panel door warm (Figure 3.1).

The cabinet itself is borne on a trolley so that it could be easily moved from one place to another. The placement of microtome inside the refrigerated cabinet ensures to maintain the uniform temperature of microtome, block holder, razor/ knife, other internal accessories, tissue, *etc.* during the operation hours for preparing cryocut sections. The microtome is manually operated by a hand driven wheel, which is fitted externally on the right hand side of the cabinet and also provided with a locking device.

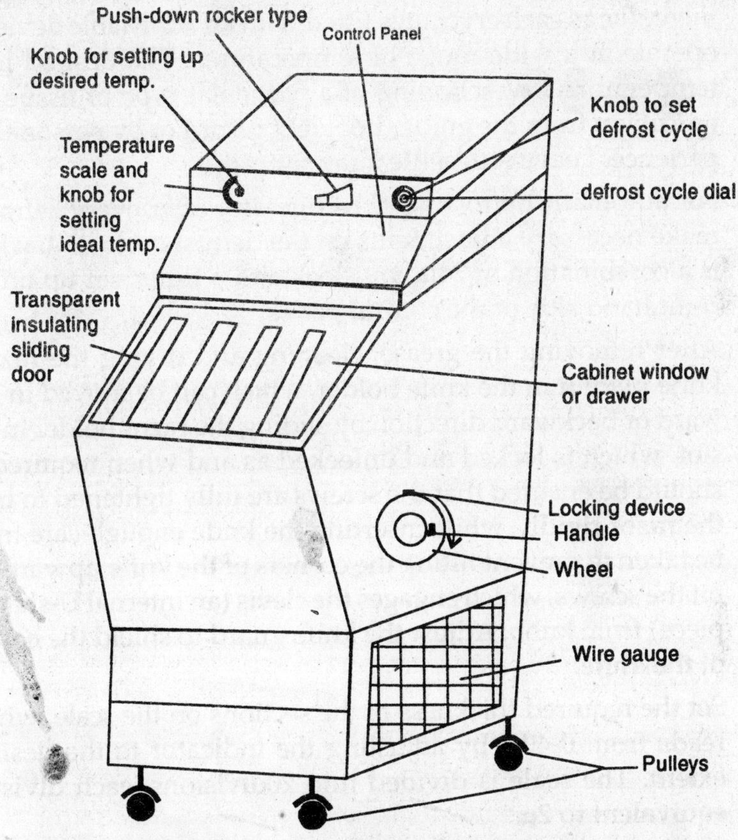

Fig. 3.1. Cryostat Cryocut Microtome

(ii) Setting up of cryocut cryostat for sectioning

To set up the cryocut cryostat in working condition proceed step by step as under:

(1) Simply connect the cryocut cryostat microtome with the main power supply but keep the push down switch in the centre of control panel at "Off" position.

(2) Move the knob of the temperature/dial scale fitted on the left side of the control panel, so that the arrow inscribed on it lies against the desired ideal temperature. Leave it on this position till the needle on the temperature scale does not coincide with the one already set at the desired temperature. When this temperature is achieved, the apparatus should be considered ready for sectioning. The ideal temperature for cutting sections by the microtome varies with the nature and type of the tissue. The optimum temperature is determined experimentally, as each cryocut is fitted with an adjustable device to operate at a wide range of temperatures. The desired ideal temperature for sectioning of a particular type of tissue can be known from the guide/booklet/ record or by personal experience. Leave it to settle automatically.

(3) For automatic defrosting of refrigerator after every 24hours, make necessary adjustments on the defrost cycle dial, which is a combination of 24hours clock and a timer, set up on the right hand side of the control panel.

(4) After removing the grease, cleaning and drying the razor/ knife place it in the knife holder, which can be moved in forward or backward direction by sliding the razor holder in the slot, which is locked and unlocked as and when required. It should be ensured that the screws are fully tightened to hold the razor rigidly, while inserting the knife enough care must be taken to prevent tilting the corners of the knife upwards to hit the screws, which engages the clevis (an internal U-shaped piece) from knob. Adjust the knife guard to shield the corner of the knife.

(5) Set the required thickness of the sections on the scale, which reads from 0-40μ, by adjusting the indicator to the desired extent. The scale is divided into 20divisions, each division equivalent to 2μ.

(6) Now, close the door of the cabinet and press down the rocker type switch to "On" position and wait for that much of period, till the needle on the left of the temperature dial does not coincide with the desired temperature indicated by arrow on the knob.

(iii) Mounting of tissue

This is the most tedious operation in the whole procedure. It should be effected expeditiously, and care must be taken not to allow the tissue to be warmed at any stage. Proceed step by step as under:

(1) For preparing the sections, either use the chilled blocks of the tissue pieces or remove a small piece of the tissue from the body of an experimental animal and trim the piece into a cube of proper dimension.

(2) Properly trimmed piece of tissue and mounted in the centre of object disc already placed on a heat sink on the left inner side of the cabinet. The heat sink has a high thermal conductivity and can draw heat from a section of the cabinet very rapidly because it is a large mass of cold metal. Frozen material can be stored in heat sink on object disc so that they are kept cold and used as and when required.

    To start with mounting the tissue, put a few drops of serum or saline water or distilled water on the object disc in the centre, place the trimmed piece on top of the fluid/liquid, wait for a few seconds, ensure that it has taken a firm grip and pour enough of liquid or distilled water so that on being freezed in the liquid, the tissue remains completely covered with frozen ice from all sides. This is achieved in 30 to 60seconds. The freezing process can be hastened by gently placing the heat extractor, which rapidly draws off heat from the tissue as per directions.

(3) When the liquid has frozen completely around the tissue, remove the object disc and mount its handle in the block holder of microtome and fix it rigidly by tightening the screw meant for this purpose.

(iv) Cutting of the sections

This is carried out successfully as per details given below:

(1) Adjust the angle of knife by experimentation so that its cutting edge lies at right angle to the elongated axis of tissue.

(2) Start cutting sections by operating the hand driven wheel in the clockwise direction after unlocking it, if previously locked.

(3) Use anti roll plate to avoid curling or rolling of sections. The curling can also be avoided by using camel hairbrush after enough skillful practice.

(4) Lift the sections from the horizontal blade of the razor. For obtaining a single section simply open the door to minimum, swing the anti roll plate away from the knife and bring an air dried microslide from ambient temperature of laboratory as

directed on and near the blade of knife holding section. The section will automatically stick itself to the underside of microslide. There is no need to press the slide on the section. Clean the surface of the knife with camel hairbrush. Bring the anti roll plate in its original position and operate the hand driven wheel for cutting another section. All these sections lifted on the slides can be stored in the cabinet of cryostat till further use or processing. For obtaining sections in the bulk, pick them up by a camel hairbrush from the blade of the knife and transfer them in a cavity block containing distilled water or a fixative or any other reagent as per requirement.

(5) For quickly advancing the tissue block near the edge of the knife, operate the fast feed control knob in the clockwise direction and when it has moved to its maximum capacity, the fast feed control knob be rotated in the anticlockwise direction.

(6) Switch off the cryocut when no more section cutting is required, by operating the push down switch and thereafter finally from the main power supply line.

(7) Remove the razor and after proper cleaning, drying and application of a thin superficial layer of grease, put it back in the anti roll box specially designed and meant for it.

(8) Lastly, close the door of the cabinet and leave the machine to defrost.

(v) Precautions

To achieve most satisfactory sectioning the following precautions must be strictly observed:

(1) Keep the razor in knife holder before switching on the cryocut.

(2) Keep window door closed all the time except when needed to open.

(3) Use fresh frozen blocks of tissue as they yield good results. Avoid cutting sections of the previously cooled and stored frozen tissues or blocks as they result either in fragmentation or in rolling of the sections.

(4) Avoid cutting sections of the large pieces of tissue, as they do not yield good results and are prone to the development of artifacts. Always cut sections at ideal temperature.

(5) In general sections, 8-10μ thick are quite suitable for all histochemical work. However, tissues chilled and prepared

in this way can be cut at any required thickness from 2-20μ.

(vi) Utility of Cryostat

Since the sections are cut at low temperature and particularly of fresh tissue, therefore the possibility of the loss of chemical components of the tissue is over ruled. At the same time, possibility of any alteration of substances or their displacement from the original position is greatly minimized. Similar is the case with the nature and degree of enzyme activity.

The second advantage lies in that the preparation of sections is not time-consuming. It is for this reason that the cryocut is being used in pathological laboratories, especially when biopsy of tissue is practiced.

Thirdly, the sections so produced are extremely good for fluorescence microscopy, autoradiography, and fluorescence microscopy more specifically for pathologic diagnosis when patient is still on the operation table.

A. HISTOCHEMISTRY OF CARBOHYDRATES

In true sense the histochemistry of carbohydrates means the localization of various forms of carbohydrates by applying specific colour tests either on the paraffin sections or fresh frozen sections or even the hand cut sections of the animal tissue and comparing them with control sections for confirmation.

From the viewpoint of the laboratory experiments, there are two polysaccharides, namely the glycogen and mucin, of which the presence in tissue sections could be readily demonstrated by histochemical means. Out of these two, glycogen is only naturally occurring group of polysaccharide, which remains in animal tissues in its natural position after fixation and prolonged processing. Various specific tests for different categories of carbohydrates are detailed in the following experiments.

Experiment No. 1

Histochemical demonstration of the presence of glycogen in the liver of vertebrates using Best Carmine Method (Best, 1906).

Theory: According to Gomori (1952), Best Carmine is an empirical stain for glycogen.

Solutions and Reagents

(A) Fixatives

(1) **Rossman Fixative (Lillie, 1965)**

It is prepared by mixing 45ml of saturated solution of picric acid in absolute alcohol with 5.0ml of formalin. The fixation of tissue is completed within 12-24hours. The washing is carried out with 90% ethyl alcohol.

(2) **Carnoy Fixative (Gatenby and Beams, 1950)**

Sol. A

Acetic acid	10.0ml
Ethyl alcohol	30.0ml

Sol. B

Glacial acetic acid	5.0ml
Ethyl alcohol	30.0ml
Chloroform	15.0ml

The fixation is completed in 3-6hours and washing is completed with two changes of absolute alcohol in 2-3hours. The two solutions A and B should be mixed in equal volumes shortly before use.

(3) **Picrate-Formalin Fixative**

40% Formaldehyde	5.0ml
Ethyl alcohol	28.0ml
Sodium chloride	0.9g
Picric acid	0.75g
Distilled water	50.0ml

Note: For glycogen test, 1ml of 5% acetic acid is added to the above mixture.

(B) Periodate reagent

Dissolve 200mg of periodic acid in 7.5ml of distilled water. To this add 67.5mg of crystalline sodium acetate dissolved in 17.5ml of absolute alcohol. This reagent should be prepared shortly before use.

(C) Reducing Rinse

Potassium iodide	1g

Sodium thiosulphate	1g
Distilled water	20ml

Note: When dissolved, add with constant stirring 30ml of ethyl alcohol and then 0.5ml of 2N hydrochloric acid (a solution of 20% HCl). A precipitate of sulphur may be formed; this is left to settle and supernatent used without filtering.

(D) Best Carmine Stain

A. Stock Solution

Carmine	2g
Potassium carbonate	1g
Potassium chloride	5g
Distilled water	60ml

Note: Mix all the ingredients and boil gently for 3-5minutes until the colour deepens, allow to cool and filter. Add 20ml concentrated ammonia to the filtrate. This keeps on well for upto 3months at - 4°C.

B. Staining solution

Stock solution A	12ml
Concentrated Ammonia	18ml
Methanol	18ml

(E) Differentiator

Absolute ethyl alcohol	20ml
Methanol	10ml
Distilled water	25ml

Procedure

(1) Prepare the cryocut sections of 8-10μ thickness of the fresh liver of the vertebrate as per procedural details given earlier and fix fresh frozen cryocut sections in the picrate formalin solution for 4-5minutes.
(2) Wash in 70% ethyl alcohol.
(3) Treat with periodate solution for 4-5minutes.
(4) Rinse the sections with 70% ethyl alcohol.
(5) Immerse the sections in reducing rinse for 4-5minutes.
(6) Treat the sections with up graded concentrations of alcohol and stain with Ehrich's haematoxylin for 15minutes.

(7) Wash in tap water (or 5% sodium bicarbonate solution) till sections turn 'blue'.
(8) Stain with Best's Carmine staining solution for 10-15minutes.
(9) Wash in differentiator just for only a few seconds.
(10) Wash very quickly in absolute alcohol.
(11) Clear in xylene and mount in DPX or Canada balsam or any other mounting medium.

For control, wash the sections with distilled water after step No. 2 above and treat them in a small cavity block either with human saliva (containing ptyalin) or 1% diastase solution for 2-3hours at 37°C in an incubator. After extraction, again wash these sections with distilled water and bring them in 70% alcohol and proceed further ahead along with the test sections simultaneously but separately till the end.

Results: The presence of red or deeply red stained fine granules indicate the areas where glycogen is present. This is confirmed by the results achieved from parallel run control sections, in which the corresponding areas show a negative response to the staining dye, due to the conversion of glycogen in glucose (by action of enzyme), which is removed during the washing process with distilled water.

Since the sections are treated with periodate and with reducing rinse, the results with Best's Carmine are more or less analogous to those with PAS Test, hence in serial sections, those regions or areas which are positive to the PAS Test and also stained red with Best's Carmine, must be considered to contain glycogen. The confirmation of this can be made by the extraction of sections by diastase or amylase before staining.

Experiment No. 2

Histochemical demonstration of the presence of glycogen in the liver of vertebrates using Periodic Acid-Schiff (PAS) Method (Hotchkin, 1948).

Theory: Picrate-formalin fixative is used for fixing the sections. The polysaccharide is unable to diffuse from its original site, hence the results obtained here are also very conclusive. Further, more authentic results can often be attained if 5% acetic acid is added to the fixative or if the sections are fixed in absolute alcohol. Both these methods are equally good for exact localization of carbohydrates.

SOLUTIONS AND REAGENTS

(A) Fixatives

Picrate-Formalin fixative may be prepared as per method given in experiment No. 1.

(B) Periodate solution

Dissolve 400mg of periodic acid in 15ml of distilled water. To this add 135mg of crystalline sodium acetate ($CH_3COONa.3H_2O$) dissolved in 35ml of absolute ethyl alcohol. This should be preserved in dark at 17-22°C and should be used only at this temperature. If it turns brown before use, it should be discarded.

(A quicker method of preparing this reagent is to dissolve 0.4g of periodic acid in 35ml of absolute ethanol and add 5ml of 0.2M sodium acetate (prepared by dissolving 2.7g in 100ml) and 10ml of distilled water).

(C) Reducing rinse

Dissolve 1g each of potassium iodide, sodium thiosulphate ($Na_2S_2O_3.5H_2O$) in 20ml of distilled water. When dissolved completely, add with constant stirring 30ml of ethanol and 0.5ml of 2N hydrochloric acid (20% HCl). The precipitate of sulphur will be formed. It should be allowed to settle down and the supernatant fluid be used as reducing rinse reagent.

(D) Schiff's Reagent

There are several different formulations available for the preparation of Schiff's reagent but on the basis of frequent use of some, both in laboratory and clinical pathology centre, the details of the three in order of preferential use are given below:

(a) Schiff's Reagent

Dissolve 1g basic fuchsin in 200ml of boiling distilled water. Shake well for 5minutes and allow it to cool to exactly 50°C. Filter and add 30ml of N Hydrochloric acid. Now, add 3g of potassium metabisulphite and allow it to stand for 24hours in amber coloured stoppered bottle. During this period, bisulphite decolourises the fuchsin, *i.e.* it forms the colourless SO_2-pararosaniline complex (leuco-basic fuchsin). Add 0.5 to 1g of decolourising activated charcoal and mix it with vigorous shaking for 1-2minutes. Filter rapidly (to avoid re-colourising leuco-basic fuchsin through a coarse filter paper or glass wool). The filtered solution should be clear and colourless. Store it for months in a well-stoppered dark coloured glass bottle preferably at -4°C. Before use, the temperature of reagent should be in proximity of room temperature.

(a) Preparation of 1N Hydrochloric acid for Schiff's reagent

Analar hydrochloric acid Sp. Gr. 1.18	98ml
Distilled water	911ml

(b) Schiff's Reagent (Banger and DeLamater, 1984)

Add 0.5g basic fuchsin to 200ml of distilled water and heat it gently for the complete dissolution. Allow it cool to 50°C and filter it with a coarse filter paper. Now, add 0.5ml of thionyl chloride ($SOCl_2$) to the filtrate, and after shaking for 30-60seconds allow it to stand for 10-12hours. To this, now add 2g of decolourising activated charcoal. Shake it for 1-20minutes. Allow to stand and then filter it. Finally store it in dark coloured glass stoppered bottle and keep it preferably at -4°C. Before use, the temperature of reagent should be restored to room temperature.

(b) Schiff's Reagent (Itikawa and Oguru, 1954)

Take 100ml of distilled water in a beaker, bring it to the boiling point. Now, dissolve 0.5g of basic fuchsin in it. Allow it to stand to cool down. Filter the mixture in a flask. Slowly pour the SO_2 through the filtrate by occasional shaking. Stop passing the gas when the solution has attained a clear red colour. Leave it stand for a period of an overnight in dark at room temperature. Add 0.5g of activated charcoal to decolourise the solution. Shake and filter and finally store it at -4°C in a dark coloured glass stoppered bottle.

Any one of the following procedures can be employed, depending upon convenience.

Procedure No. 1

(1) Prepare fresh cryocut sections of the tissue, preferably liver of vertebrate.

(2) Fix the sections in picrate -formalin fixative for 4-5minutes. Alternatively, the sections can also be fixed in absolute alcohol or in picrate-formalin solution to which 3% solution of acetic acid has been added.

(3) Wash the sections with 70% ethyl alcohol.

(4) Treat the sections with periodate solution for 4-5minutes at

room temperature in such a way that the sections remain completely submerged in the solution.

(5) Again wash the sections in 70% ethyl alcohol.

(6) Immerse the sections in reducing rinse for 5minutes. (Before proceeding ahead it should be fully ensured that the last trace of periodate has been completely removed and for this either one or more change of reducing rinse be given or duration of treatment should be extended beyond 5minutes).

(7) Wash with 70% alcohol.

(8) Immerse in Schiff's reagent for 30minutes.

(9) Wash with three changes of SO_2 water.

(10) Dehydrate in ascending grades of alcohol and mount in DPX mounting medium.

Results: The development or appearance of red or purple red colour will indicate the presence of glycogen or PAS positive reaction.

For confirmation also perform the control test as under

After washing with 70% alcohol at step No. 3, rinse the sections with distilled water. Treat the sections with human saliva and again wash with distilled water. Bring the sections in 70% alcohol and proceed ahead from step No. 4 onwards along with the test for reducing, staining, dehydration, clearing and mounting. In this control there will be either no accumulation of red dye and if at all it will be much less. From this, it can be concluded that the salivary enzyme ptyalin has reacted with glycogen, converting it into glucose, hence development of red colour in corresponding areas, as in test sections.

Preparation of SO_2 water.

1N Hydrochloric acid	25ml
0.5% aqueous sol. of potassium metabisulphite	50ml

Mix immediately before use.

In case of above experiment where nuclei have to be stained, the following procedure with slight modifications as under should be adopted.

ADDITIONAL STAINS REQUIRED

Celestin Blue Solution

This is prepared by dissolving 2.5g of iron alum in 50ml of distilled water and allowing it to stand for a period of 12hours. To this add 0.25g of Celestin Blue B stain and boil it for 3-4minutes. Allow it to stand to cool and then filter it and add 7ml of glycerine.

Orange G solution

Dissolve 2g of orange G in 100ml of 5% aqueous solution of phosphotungstic acid and allow it to stand for 24hours and use only the supernatant fluid when required.

Procedure No. 2

(1) Prepare fresh frozen sections.

(2) Rinse with 70% alcohol.

(3) Treat with periodate solution for 5minutes at room temperature.

(4) Rinse with 70% alcohol.

(5) Treat with reducing solution for 1minute at room temperature.

(6) Rinse with 70% alcohol.

(7) Immerse the sections in Schiff's reagent for 20minutes.

(8) Wash with running water for 10minutes.

(9) Stain nuclei with Celestin Blue solution for 2-3minutes (followed by staining with Mayer's haematoxylin for 2-3minutes).

(10) Differentiate in 1% acid alcohol.

(11) Wash with running tap water for 30minutes.

(12) Counter stain the sections with Orange G solution for 10seconds.

(13) Wash with distilled water for 30seconds or till the sections become pale yellow.

(14) Dehydrate in ascending grades of alcohol, clear in xylene and mount in Canada balsum or DPX.

Results: Development of deep purplish colour will indicate the presence of hexose containing substances.

Experiment No. 3

Histochemical demonstration of the presence of glycogen in the liver of vertebrates using Periodic Acid-Schiff (PAS) Method of McManus (1946).

Procedure

This procedure is applicable both for the fixed frozen sections and the fresh frozen sections. However, the following details are exclusively meant for the fresh frozen sections.

(1) Straight away oxidize the sections in 1% aqueous solution of periodic acid for 10minutes.

(2) Wash in running tap water for 4-5minutes.

(3) Stain with Schiff's reagent for 8-10minutes.

(4) Re-wash in running tap water for 5minutes.

(5) Counter stain nuclei with Celestin Blue haemotoxylin.

(6) Dehydrate in ascending grades of alcohol.

(7) Clear in xylene and mount in Canada balsum or DPX.

Results: Appearance of magenta coloured fine granules will indicate the presence of glycogen and other periodate reactive carbohydrates, while the nuclei will stain blue.

Experiment No. 4

Histochemical detection of the presence of mucin and acidic mucopolysaccharides in the animal tissue using Alcian Blue Method.

Theory: This is an extremely valuable stain for 'mucin' and acidic mucopolysaccharides.

Solutions required

(a) **0.1% Alcian blue solution**

0.1% solution of Alcian blue is prepared by dissolving 0.1g of Alcian blue in 100ml of 3% acetic acid. If Alcian blue is not available then use Alcian green 3BX or Alcian green 2GS but the concentration of this solution should be increased upto 1% and pH maintained at pH 2.5.

(b) **0.1% aqueous Alcian blue solution**

0.1% solution of Alcian blue is prepared by dissolving 1g of Alcian blue in 100ml of distilled water. If this solution is saturated with thymol, it can be stored for weeks together but it should never be used before filtration.

Procedure

This procedure is applicable both for the fixed frozen sections and the fresh frozen sections. However, the procedural details given here are exclusively meant for the fresh frozen sections.

(1) Fix fresh frozen cryocut sections in picrate-formalin fixative for 4-5minutes in a small cavity block.

(2) Pass the sections through degraded concentration of alcohol to distilled water.

(3) Remove water and stain the sections in 0.1% Alcian blue solution for 15-20minutes.

(4) Re-stain in 0.1% of Alcian blue solution for 2-3minutes.

(5) Wash well in tap water.

(6) Dehydrate and mount in Canada balsum or DPX.

Results: Mucin and acid mucopolysaccharides are stained blue or blue green while nuclei are rendered red.

Experiment No. 5

Histochemical detection of mucin and acidic mucopolysaccharides in liver using Hales Colloidal Iron Method.

Solutions required

(a) Dialyzed iron

Pharmaceutical grade of dialyzed iron (5% Fe_2O_3)	1vol.
2N Acetic acid	1vol.

(b) Acidic ferrocyanide solution

It is prepared by mixing 85ml of 1% solution of potassium ferrocyanide with 15ml of 1N HCl.

Procedure

(1) Prepare cryocut sections (15-20μ) of the liver of frog or rat or any other animal.

(2) Fix them in picrate-formalin for 5-10minutes.

(3) Bring the sections in water via descending grades of alcohol.

(4) Keep the sections completely submerged in a horizontal position in the dialyzed iron solution for 10-12minutes.

(5) Wash the sections with distilled water.

(6) Immerse the sections in acidic ferrocyanide solution for 10-15minutes.

(7) Re-wash with distilled water (The sections can be counter stained with Safranin at this stage).
(8) Dehydrate with ascending grades of alcohol.
(9) Clear in xylol and mount in Canada balsam or DPX.

Results: Sites containing either mucin or acidic mucopolysaccharides or both are stained blue.

Experiment No. 6

Histochemical detection of glycogen and starch in the liver of frog, rat, *etc.* using Lugol's Iodine Test.

Solution required

Lugol's Iodine

Iodine	1g
Potassium iodide	2g
Distilled water	100ml

Procedure

(1) Prepare cryocut sections (10-15μ) of a piece of liver.
(2) Fix in picrate-formalin fixative for 4-5minutes.
(3) Wash the sections in 70% alcohol.
(4) Stain for variable periods with Lugol's Iodine.
(5) Rinse in distilled water at least twice or thrice to remove excess of stain.
(6) Dehydrate in ascending grades of alcohol.
(7) Clear in xylol and mount in Canada balsam or DPX.

Results: Sites containing glycogen are stained brownish whereas starch is coloured deep blue.

For the confirmation of the above results, it is advisable to run a control along with test sections after step No. 3 above.

Preparation of Control Sections

After step No. 3, rinse the sections in distilled water and treat them for 3hours at 37°C with human saliva or 5% solution of α-amylase or 1% solution of diastase depending on the availability. After this treatment all glycogen and starch will be removed. Rinse the sections in distilled water and proceed ahead with staining and other steps along with test sections but in separate cavity block, till the end.

Results: The sites corresponding to positive test in test sections will remain unstained in control section, thus confirming that positive response of sites in test was on account of glycogen or starch.

Experiment No. 7

Histochemical detection of chitin using Campbell's Chitoson Test.

Procedure

(1) Prepare cryocut sections (15-20μ) of the exoskeleton of some insects, for instance, the newly moulted cockroaches, soft hardened cuticle of blow fly larvae, or housefly maggots or caterpillars of butterfly or moth, *etc.*

(2) Store these sections in 10% formalin or in distilled water, (if not required for immediate use).

(3) Examine one or two sections on a slide under the microscope and visualize different zones, such as, the epicuticle, procuticle (exocuticle and endocuticle) and note their colours.

(4) Now, treat the sections, when still on slide with 2% iodine solution. If formalin fixed sections are to be used, then first of all they should be thoroughly washed with distilled water before being treated with iodine solution. If a brown colour is developed in the zone of procuticle, it should be regarded as an indication for the presence of chitin.

(5) This is further confirmed when iodine solution is replaced with 1% sulphuric acid on the same section. The appearance of a violet, reddish violet or black colour should be regarded as a confirmatory test for the presence of chitin.

This test can also be applied on hand cut sections of unhardened or untanned insect or larval cuticle. The sections can be prepared by a sharp razor using pith or simply by sharp scissors depending on the skillful practice of the operator.

B. HISTOCHEMISTRY OF LIPIDS

The term lipid comprises of all those substances that are extracted from animal tissues by dissolving them in fat solvents, such as chloroform, carbon tetrachloride, benzene, acetone, ether, *etc.* For a histochemist, the term lipid is synonymous with fat.

Lipids are one of the most important metabolites because they are largely stored as reserve food material and along with carbohydrates they act as a great source of biological fuel and play a very important role in metabolism.

Histochemically demonstrable lipids are differentiated into: (a) simple lipids, (b) compound lipids, (c) derived lipids, (d) sterols and sterol esters, and (e) allied substances.

Histochemical tests of lipids fall into two categories:

(1) Test involving physical methods, which depend on the readiness of lipophilic molecules to partition between their solvent and the lipid in the section. These are, therefore, tests for the degree of fattiness of lipid material.

(2) Tests involving chemical methods, such as Sudan black B, Nile Blue, Acid Haematin methods, *etc.*

Experiment No. 1

Histochemical characterization of lipids in the animal tissue using Sudan black B method.

Theory: Sudan black B is a fat-soluble colouring stain. It is soluble in absolute alcohol but insoluble in water. It is used as a saturated solution in 70% ethanol and is very much soluble in 50% alcohol. Because of its intense colour and strong affinity for all or most classes of lipids, the colouration of the tissue components by Sudan black B is often taken as the histochemical criterion for the presence of lipids. A simple draw back of this procedure is the fact that the section is exposed to 70% alcohol, which may itself affect the lipids but for the routine laboratory experiments, it hardly makes any difference.

Procedure

(1) Prepare cryocut sections (15-20μ) of the fixed tissue (10% formalin) or the fresh frozen sections (15-20μ) of unfixed fresh tissue, for example, the arthropod cuticle or ovarioles or the fat bodies of cockroach (The ovarioles as such without sectioning can also be used).

(2) Transfer the sections in a small cavity block containing water. If the sections are those of fixed tissue, wash them thoroughly well with distilled water otherwise proceed ahead.

(3) Remove water and give two changes of 70% alcohol each time for 2-3minutes with occasional agitation by means of pipette.

(4) Remove 70% alcohol and stain with freshly filtered solution of Sudan black B and wait for 4-5minutes.

(5) Remove the stain and immediately rinse the sections at least twice with 70% alcohol to remove the excess of stain as well as any deposit of the dye. A mixture of glycerine and water; 85:15 v/v can be used as differentiator in place of 70% alcohol. Take care that the differentiation process is completed in minimum possible time else the stain from inside the tissue after staining may be removed.

(6) Bring down the sections in 50% alcohol.

(7) Replace 50% alcohol with distilled water.

(8) Mount the sections on a slide under the cover slip in a drop of glycerine for temporary preparation or for permanent mount use DPX mounting medium.

(9) Examine your preparation under microscope.

Results: The sites stained blue black with Sudan black B indicate the presence of lipids.

In the stained preparation of the cuticle, the layer of epicuticle is stained deep blue black or grey whereas endocuticle remains either unstained or is stained very faintly.

Experiment No. 2

Histochemical localization of lipids in the nervous tissue of animals using Sudan black B.

Procedure

Prepare a thin smear of the grey matter from the spinal cord either of a freshly slaughtered fowl or goat on a microslide and stain the smear by following the procedure of Experiment No. 1.

Result: The nerve cells are stained blue black or grey.

Sudan black B stain, after being used should be restored for use in future. Some workers prefer to use 0.7% Sudan black B in propylene glycol and to differentiate in 85% propylene glycol.

Experiment No. 3

Detection of lipids in animal tissues using Oil Red O Method.

Theory: Oil Red O is a strong colouring dye of fats but requires a high concentration of fatty material in sections.

Solutions required

(a) **Stock solution**

Prepare a saturated solution of Oil Red O in 99% isopropyl alcohol (Approx. 0.5g of ORO). At the time of staining mix 3 volumes of this solution with 2 volumes of distilled water.

(b) **Staining solution**

Prepare by mixing 3 volumes of solution (a) and 2 volumes of 1% solution of dextrin, which prevents the precipitation of dye. This solution should always be filtered shortly before use.

Procedure

(1) Prepare fresh frozen cryocut sections of the animal tissue containing high percentage of fats, e.g. the ovarioles of cockroach, fat bodies of insects, *etc.*

(2) Transfer the sections in a cavity block and fix them in 10% formalin solution or calcium-formal fixative for 2-3 minutes.

(3) Rinse the sections in 60% isopropyl alcohol.

(4) Stain the sections by immersing them in Oil Red O staining solution for 10-15 minutes.

(5) Remove excessive stain by differentiating with 60% isopropyl alcohol.

(6) Briefly rinse with distilled water.

(7) Wash with water.

(8) Treat the sections with 5% sodium bicarbonate for a few minutes.

Result: Deeply stained areas will indicate the presence of lipids.

Experiment No. 4

Histochemical detection of lipids in animal tissues using Nile Blue Method.

Theory: An aqueous solution of Nile Blue A contains the blue cations of the dye; the red spontaneous oxidation product of it, are namely Nile red; and the orange red amino bases. Nile red dissolves in all fatty lipid components and acts as a general lipid colourant.

Solutions required

(a) **1% Nile Blue**

Nile Blue	1g
Distilled water	100ml

(b) **0.02% Nile Blue**

Nile Blue	20mg
Distilled water	100ml

Procedure

(1) Prepare cryocut sections.

(2) Warm all solutions separately to either 37°C or 60°C before starting the reaction.

(3) Immerse the sections in a 1% aqueous solution of Nile Blue at 37°C or 60°C for 4-5 minutes.

(4) Wash quickly with warm water.

(5) Immerse quickly in warm 1% acetic acid at 37°C or 60°C for 30 seconds.

(6) Wash with warm water and mount in glycerine.

Compare this section with a second section, which is stained and differentiated the same way as in first case. Re-stain with a 0.02% aqueous solution of Nile Blue at 37°C or 60°C for 4-5 minutes. Then wash quickly in warm water; differentiate in warm 1% acetic acid for 30seconds; wash with warm water and mount in glycerine.

Results: The sites containing lipids (hydrophobic) are stained pink, free fatty acids are stained pink to blue and phospholipids are stained blue.

Experiment No. 5

Histochemical detection of lipids in the animal tissues using Acid Haematein Method.

Solution required

0.1% Haematoxylin	100ml
1% Sodium periodate	2.0ml

Boil the mixture, allow it to cool and finally add 2ml of glacial acetic acid. This solution should always be prepared afresh.

Procedure

(1) Prepare cryocut sections of the tissue and fix them in formalin-calcium fixative (prepared by mixing equal volumes of 10% formalin and 10% calcium chloride solutions. To the mix add a piece of chalk to maintain pH at neutrality).

(2) Treat sections with 5% potassium dichromate containing 1% calcium chloride at room temperature for 16-18hours.

(3) Now, continue the same treatment for a period of further 2-3 hours at 37°C.

(4) Wash the sections with distilled water (two changes) for 20-30 minutes.

(5) Stain the sections with freshly prepared staining solution at 37°C for about 2 hours.

(6) Wash with distilled water at 37°C.

(7) Differentiate in an aqueous solution containing equal volume of 0.25% sodium tetraborate and 0.25% potassium ferrocyanide at 37°C for 60 minutes.

(8) Counter stain the nuclei with 1% methyl green.

(9) Wash with warm water at 37°C.

(10) Mount the sections in glycerine jelly.

Results: Lecithin and sphingomyelin are stained blue black.

C. HISTOCHEMISTRY OF PROTEIN

The proteins form one of the most important and complex groups of biological molecules comprising the chief nitrogenous constituents of the body tissues and of our food. The proteins are built up exclusively of amino acids. Therefore, on their hydrolysis no other product, except the amino acids, are yielded.

The proteins are differentiated into three major classes: namely the basic (positive to fast green at high pH), acidic (positive to fast green and toluidine at low pH) and neutral proteins.

Since proteins are invariably distributed in every cell and tissue, hence their localization is not very specific. However, there are certain classical tests for proteins, which are specifically associated with certain amino acids comprising the type of proteins.

Experiment No. 1

Histochemical localization of proteins in the animal tissues using

Mercury-Bromophenol Blue Method.

Solutions required

(a) 1% Bromophenol Blue in alcohol already saturated with mercuric chloride.

(b) 1% mercuric chloride and 0.05% bromophenol blue in 2% acetic acid.

Procedure

(1) Prepare cryocut sections and fix in any good fixative preferably 10% formalin or Carnoy's fluid in a cavity block.

(2) Wash the sections with distilled water to remove the fixatives.

(3) Remove water and stain either with staining solution (a) or (b) for about 1 ½ -2 hours at room temperature.

(4) Rinse the sections with 0.5% acetic acid for 4-5 minutes (at least 2 changes).

(5) Remove acetic acid and dehydrate directly with tertiary butyl alcohol or with ascending grades of alcohol.

(6) Clear in xylene.

(7) Mount in Canada balsam or DPX.

Results: Sites containing proteins are stained blue. This is most specific for proteins and peptides.

Experiment No. 2

Histochemical localization of proteins in the animal tissues using Mercuric-Bromophenol Method (Chapman, 1975).

Solutions required

(a) **Staining Solution**

Mercuric chloride ($HgCl_2$)	1.0g
Sodium bromophenol blue	0.05g
2% aqueous acetic acid	100.0ml

(b) 0.5% aqueous acetic acid solution.

Procedure

(1) Prepare fresh frozen cryocut sections (8-10µ) of the tissue.

(2) Fix in any general fixative preferably 10% formalin or Carnoy's fluid for 4-5 minutes.

(3) Wash the sections with at least two changes of distilled water.

(4) Stain the sections with bromophenol staining solution (a) for 12-15 minutes.

(5) Rinse the sections with at least two changes of 0.5% aqueous acetic acid solution (b) for 15-20 minutes.

(6) Blot and dehydrate first with 90% alcohol and then with absolute alcohol, in each case at least with two changes.

(7) Replace alcohol with xylene and wait for 2 minutes.

(8) Then treat with a mixture of 100ml of xylene and 0.5ml of n-butyl amine until sections turn blue.

(9) Finally clear in two changes of xylene and mount in Canada balsam or DPX.

Results: Proteins and peptides that are not washed away during processing are stained blue.

Experiment No. 3

Histochemical localization of proteins in the animal tissues using Million's Test (Gomori, 1952).

Solutions required

(a)	Mercuric acetate	5g
	Trichloroacetic acid (TCA)	15g
	Distilled water	100ml
(b)	1% aqueous solution of sodium nitrate	

Procedure

(1) Prepare fresh frozen cryocut sections (8-10μ) of the animal tissue and fix in any suitable fixative, preferably in 10% formalin or picrate formal or Carnoy's fluid, *etc.*

(2) Transfer them in a cavity block and wash them well with distilled water. The complete removal of the fixative should be ascertained before proceeding ahead.

(3) Remove water and incubate the sections in solution (a) at 37°C in an incubator for 7-10 minutes.

(4) Now, to the above incubation medium in the same cavity block (Step 3) add solution (b) and again leave it for incubation at 37°C for 25-80 minutes.

(5) Quickly rinse the sections with 70% alcohol, and dehydrate the sections by passing through 90% and absolute alcohols.

(6) Clear in xylene and mount in Canada balsam or DPX.

Results: Sites of tissue containing tyrosine are rendered pink to brick red in colour.

Experiment No. 4

Histochemical detection of proteins in animal tissues using Baker's method (Baker, 1956).

Theory: The phenolic amino acid, tyrosine is a good marker for most proteins, except collagen. The most widely practised histochemical technique for this amino acid has been simple modification of the Million's Test (Bensley and Gersh, 1933). Baker developed the following technique (from Biochemical Folin reaction), which yields a characteristic stain in 10μ-15μ thick sections. In this reaction, the phenol is converted into a nitrosophenol to form a red compound, which is readily discernible in microscopic examinations of the sections.

Solution required

(a) **Mercuric sulphate solution**

Take 90ml of distilled water in a beaker and add 10ml of A-R (analytical reagent) grade of concentrated sulphuric acid to it. Then add 10g of A-R grade of mercuric sulphate. Heat the mixture to ensure complete mixing of all. Allow it to cool and then make up the volume to 200ml by addition of distilled water. This solution is quite stable and can be stored at low temperature.

(b) **Sodium nitrite solution**

Dissolve 0.25g of sodium nitrite (AR grade) in 100ml of distilled water with occasional stirring to ensure complete dissolution.

(c) **Acetic-alcohol fixative**

5% glacial acetic acid in absolute ethyl alcohol.

Procedure

(1) Prepare fresh cryocut sections of the tissue and transfer them to small cavity block.

(2) Fix the sections in acetic-alcohol fixative (c) for 2-10 minutes.

(3) Completely re-hydrate the sections by passing them through descending grades of alcohol.

(4) Mix 25ml of mercuric sulphate solution (a) with 2.5ml of sodium nitrite solution (b) in a 50ml beaker and stir well.

(5) Now, transfer the sections from step No. 3 in the beaker containing mixture prepared at step 4 and gently heat it to just boiling point.

(6) Allow the beaker to stand for sometime and then take out the sections again in a cavity block.

(7) Wash the sections thoroughly well with at least three changes of distilled water.

(8) Either directly mount in glycerine jelly for immediate microscopic examination or dehydrate the sections in ascending grades of alcohol and after clearing in xylene mount in Canada balsam or DPX for permanent preparation to be examined afterwards.

Results: Most proteins are stained red or pink or yellowish-red.

Experiment No. 5

Histochemical detection of proteins in animal tissues using Ninhydrin-Schiff reaction (Yasuma and Ichikawa, 1953).

Solutions required

(a) **Ninhydrin solution**

Ninhydrin	5.0g
Absolute alcohol	100.0ml

(b) **Schiff reagent**

In a beaker take 100ml of boiling distilled water and dissolve 0.5g of basic fuchsin in it. Mix the two by vigorous shaking for about 4-5 minutes. Allow it to stand and cool down to 50°C. Filter and add 15ml of 1N Hydrochloric acid, cool further to 25°C and add 1.5g of sodium or potassium metabisulphite. This solution is allowed to stand for 24 hours in dark coloured glass stoppered bottle. During this period, the bisulphite decolourises the fuchsin, *i.e.* it forms the colourless SO_2-pararosanline complex (leuco-basic fuchsin). Impurities and other coloured products are then removed by adsorption on to activated charcoal. Therefore, add 0.25 to 0.5g of a decolourising activated charcoal to this solution and shake well for about 1-2 minutes. Filter it rapidly (to avoid recolourisation of the leuco-basic fuchsin through a coarse filter paper). Finally filtered solution should be clear and colourless. It can be stored for prolonged use in a well-stoppered dark coloured glass bottle preferably at -4°C.

It is advisable to use the reagent only when it has attained the room temperature.

(c) **Bisulphite solution**

Mix 5ml of 10% potassium or sodium metabisulphate, 5ml of N HCl and 90ml of distilled water.

Procedure

(1) Prepare cryocut sections (10-15μ) of fixed or fresh frozen tissue. Transfer them in a cavity block.

(2) If fixed frozen sections, then dewax with xylol or otherwise bring them to absolute alcohol. To ensure the complete dehydration, the treatment with alcohol should be repeated twice each after 2 minutes.

(3) Transfer the sections in ninhydrin solution (a) and incubate at 37°C in a bath for 10-20hours.

(4) After incubation, thoroughly wash the sections with tap water for 3-5 minutes.

(5) Place in 1N HCl (8.5ml of concentrated HCl + 91.5ml distilled water) at room temperature for 1minute.

(6) Place in 1N HCl at 60°C.

(7) Rinse in 1N HCl at room temperature for 1minute.

(8) Treat with Schiff reagent for 30-60 minutes.

(9) Remove the reagent and rinse in three changes of freshly prepared bisulphite solution, each time for 2 minutes.

(10) Rinse in distilled water and if desired, counter stain with 1% light green stain for 30-60 seconds.

(11) Dehydrate, clear in xylol and mount in Canada balsam or DPX.

Results: The sites containing proteins and peptides, except arginine rich protamines, are stained red or purple red.

Experiment No. 6

Histochemical detection of proteins containing arginine in animal tissues using Sakaguchi reaction (Sakaguchi, 1925 and Deitch, 1961).

Theory: In this reaction α-naphthol is linked to the quanido-amino group of arginine at a particular alkaline pH. Colour is then produced by the addition of hypobromite or hypochlorite. Other investigators using the dichloro-naphthol and hypomite demonstrated the rapid development of a stronger colour reaction.

Solutions required

(a) **Lewitsky's fluid (fixative)**

10% formalin (*i.e.* 10% of the 40% formaldehyde)	1vol.
1% chromic acid (chromium trioxide)	1vol.

Mix the two just before use. It is necessary to wash well after this fixative. For this, the section should be washed at least with 3 changes of distilled water each time for 5 minutes.

(b) **Celloidin solution**

Ether (diethyl ether)	50ml
Absolute alcohol	50ml
Celloidin	1g

This is just to protect and preserve the sections from washing away from slides.

(c) **Reaction medium**

1% aqueous solution of sodium hydroxide	30ml
1% solution of 2,4 dichloro-α-naphthol in 70% ethanol	0.6ml
sodium hypochlorite	1.2ml

Mix these immediately before use.

(d) **Urea solution**

5% aqueous solution

(e) **Sodium hydroxide solution**

1% aqueous solution

(f) **Alkaline glycerol**

Glycerol	9ml
10% aqueous solution of sodium hydroxide	1ml

(g) **Mounting medium**

Aniline	0.1ml
Xylene	100ml

Add one volume of this to four volumes of permanent mounting medium (routinely used)

Procedure

(1) Prepare fresh frozen cryocut sections of the tissue.

(2) Fix them either in Lewitsky's fluid for 15 minutes or in 10% neutral formalin The section of tissues already fixed in either fixative, and prepared after embedding in paraffin following routine procedure can also be used.

(3) Bring the sections in absolute alcohol by passing through successive ascending grades of alcohol.

(4) If sections have already been placed on slide, dip the slide in celloidin solution to fix them firmly. Drain off the excessive celloidin solution and allow the slide to dry in air.

(5) Bring the sections down to distilled water by passing the sections through successive descending grades of alcohol.

(6) Treat the sections with freshly prepared reaction mixture (e) for 5-6 minutes.

(7) Rapidly rinse the sections in 5% solution of urea. This is just meant for stopping the reaction. Therefore, its prolonged treatment should be avoided.

(8) Place the sections in 1% solution of sodium hydroxide for 4-5 minutes.

(9) For temporary preparations, mount in alkaline glycerol (f) or for permanent mount dehydrate the sections in tertiary butyl alcohol, then into aniline oil for 3 minutes, wash in xylene for 10seconds and make the permanent preparation by mounting in mounting medium.

Results: The sites taking up orange red colour indicate the presence of arginine.

Experiment No. 7

Histochemical detection of basic proteins in animal tissues using Fast Green Method (Geschwind, 1953).

Solutions required

(a) **Trichloracetic acid** 5% in distilled water

(b) **Fats Green**

0.1% in citrate-phosphate buffer (pH 8.0).

(c) **Buffer solution (McIlvaine's)**

0.1M citric acid 2.75ml

(dissolve 0.48g in 25ml distilled water),

0.2M disodium phosphate 97.25ml

(dissolve 7.1g in 250ml distilled water)

Procedure

(1) Prepare fresh cryocut sections.

(2) Fix them in acetic alcohol (5:95 v/v).

(3) Bring in distilled water by passing through successive descending grades of alcohol.

(4) Immerse sections for 15 minutes in 5%TCA, the temperature of which should be around 95°C (This can be achieved by putting a coupling jar in a beaker of boiling water).

(5) Wash in three changes of 70% alcohol, each time for 10 minutes and then in distilled water to remove all traces of TCA.

(6) Stain in Fast green (b) for 30 minutes at room temperature.

(7) Rinse with distilled water.

(8) Directly transfer the sections to 95% alcohol.

(9) Dehydrate, clear and mount in Canada balsum or DPX.

Results: Basic protein stains green or greenish-blue, histones are coloured deep blue

D. HISTOCHEMISTRY OF NUCLEIC ACID

Nucleic acids are long chain polymers, built by a large number of units, called nucleotides. Nucleic acid on hydrolysis yields phosphoric acid, a sugar, pyrimidines and purines.

In general, the two kinds of nucleic acids are universally recognized depending upon the sugar present in the molecule, one is called ribose nucleic acid (RNA) and the other deoxyribose nucleic acid (DNA).

Experiment No. 1

Histochemical detection of deoxyribo nucleic acid in animal tissues using Fuelgen reaction (Fuelgen and Rossenbeck, 1924).

Theory: Neither DNA nor RNA contains free aldehyde groups. The demonstration of nucleic acid depends on the reaction of stain with phosphate moiety and the production of aldehydes from deoxyribose sugar. However, acid hydrolysis can unmask the aldehyde groups in the deoxyribose moiety of DNA but not in RNA.

Solutions required

(a) **Schiff Reagent**

Procedure for preparing this reagent remains the same as described in experiment no. 5 under proteins.

(b) **SO_2 water**

1N HCl	5ml
10% aqueous sol. of potassium metabisulphite	15ml
Distilled water	90ml

(c) **1N hydrochloric acid**

AR grade concentrated HCl (Sp. Gr. 1.18)	8.5ml
Distilled water	91.5ml

Always add acid to water.

Procedure for Test Sections

(1) Prepare cryocut sections (10-15μ) of fresh frozen tissue.

(2) Fix them in acetic ethanol (1:3, v/v) for 10 minutes.

(3) Bring the sections in water by passing through successive descending grades of alcohol.

(4) Rinse in 1N HCl at room temperature for 15 minutes.

(5) Treat the sections in 1N hydrochloric acid at 60°C for a few minutes (approx. 5-6 minutes; this treatment unmask the aldehyde groups in DNA but not in RNA).

(6) Transfer the sections in 1N HCl at room temperature for 1 minute.

(7) Treat the sections with Schiff reagent in the dark for 30-60 minutes.

(8) Drain the reagent and rinse the sections with 3 changes of freshly prepared SO_2 water (b) each time for 2 minutes.

(9) Rinse in distilled water and counter stain (if required) in 1% light green for 1minute.

(10) Dehydrate, clear and mount in Canada balsam or DPX.

Results: The sites pertaining to DNA will stain magenta colour.

For confirmation of the above results, a set of control section should be run parallel to test section as under:

Procedure for Control

(1) Prepare cryocut sections (10-15μ) of the fresh frozen tissue.
(2) Fix them in acetic ethanol for 10 minutes.
(3) Dehydrate the sections in different grades of alcohol.
(4) Rinse in 1N HCl at room temperature for 15 minutes.
(5) Treat with Schiff"s reagent in the dark for 30-60 minutes.
(6) Drain the reagent and rinse thrice with SO_2 water.
(7) Rinse in distilled water and counter stain with 1% light green for 1minute.
(8) Dehydrate, clear and mount in Canada balsam or DPX.

Any stain produced in the control sections should not be regarded due to DNA but it must be substracted from the result obtained by the full test procedure.

Experiment No. 2

Histochemical detection of DNA in animal tissues using Pyronin-Methyl green Method (Elias, 1969).

Solution required

Methyl green	0.5g
Acetate buffer (pH 4.1)	100.0ml
Pyronin GS or Y	0.2g

Procedure

(1) Prepare cryocut sections either of fixed tissue or fresh frozen material (preferably Carnoy's fluid should be used as fixative).
(2) If fixed cryocut sections are used then dewax and bring them to distilled water or if frozen then straight away proceed ahead.
(3) Stain the sections at 37°C for one hour.
(4) Wash the sections quickly with ice-cold distilled water approx. 1-2seconds.
(5) Wipe off the water with blotting paper.
(6) Rinse with vigorous agitation in tertiary butyl alcohol.
(7) Finally dehydrate by giving 2 changes of tertiary butyl alcohol, each time for 5 minutes.

(8) Clear and mount.

Results: DNA sites will appear green to blue while RNA red.

Differentiation with ice-cold water is very important for satisfactory results. For confirmation, it is advisable to run a control slide parallel to test by following the under mentioned extraction method. For this RNA should be removed.

A. REMOVAL OF RNA FROM CONTROL SECTIONS BY EXTRACTION METHOD (BRACHET, 1940)

Procedure

(1) Prepare cryocut sections preferably in Carnoy's fluid.

(2) Bring both the test and control sections in water.

(3) Place test sections in ribonuclease solution and control sections in distilled water and incubate at 37°C for 1hour.

(4) Wash both sections in distilled water and proceed further ahead as in general test procedure of pyronin-methyl green method.

Result

(1) In test sections RNA sites will be negative and DNA sites green.

(2) In control sections RNA takes red and DNA the green colour.

B. REMOVAL OF DNA FROM CONTROL SECTIONS BY EXTRACTION METHOD (BRACHET, 1940)

Solution required

(a)	deoxyribonuclease	10.0mg
	0.2M Tris buffer(pH 7.6)	10.0ml
	Distilled water	50.0ml

(1) Prepare cryocut sections and fix preferably in Carnoy's fluid.

(2) Wash the sections and dehydrate both the test and control sections.

(3) Treat test with deoxyribonuclease solution and control sections in Tris buffer pH 7.6 at 37°C for 4hours in an incubator.

(4) Wash in running tap water and proceed ahead for staining using Fuelgen method.

Result

(1) In test sections DNA sites are negative.

(2) In control sections DNA gives red.

Experiment No. 3

Histochemical localization of RNA in animal tissues using hydrochloric acid method followed by Pyronin-Methyl green staining.

Theory: The method is based on the fact that 1N hydrochloric acid at 60°C extracts entire RNA from sections in 5 minutes, it also causes sufficient alteration in DNA to make it stained with pyronin and not with methyl green component of methyl green pyronin mixture.

Solution required

Methyl green	0.5g
Acetate buffer (pH 4.1)	100.0ml
Pyronin GS or Y	0.2g

Procedure

(1) Proceed with two cryocut sections of fresh frozen tissue.

(2) Fix both sections in acetic-alcohol (either 1:3 or 1:5 v/v as required).

(3) Take both the sections through graded concentrations of alcohol to 0.2M acetate buffer at pH 4.2.

(4) Immerse one section in 1N hydrochloric acid, which is at 60°C and maintain at 60°C for 5 minutes.

(5) Wash both the sections with 0.2M acetate buffer.

(6) Stain both in methyl green pyronin mixture.

(7) Blot dry, immerse in n-butyl alcohol for 2 minutes.

(8) Mount in Canada balsam or DPX.

Result: RNA sites are stained red in sections not treated with HCl but are totally absent from acid hydrolyzed sections. DNA, on the other hand, stains green-blue in non-hydrolyzed but reddish or purple in hydrolyzed sections.

If any moisture is left, methyl green will be lost from the sections at next step.

Experiment No. 4

Histochemical detection of DNA in animal tissue using Kurnick's Methyl Green Method.

Theory

The phosphate group of DNA is not freely available in cells. These are usually linked to basic groups, such as arginine and lysine moieties of nuclear histones. Therefore, to unmask them, some acidic treatment is required to remove the phosphate group available to the dye. This method thus involves treatment of sections with 0.1N hydrochloric acid to split DNA from histone-like proteins.

Solutions required

(a) 0.2M acetate buffer as in pyronin methyl green method

(b) Staining solution

Dissolve 0.25g of methyl green in 100ml of 0.2M acetate buffer at pH 4.2.

Procedure

(1) Prepare fresh cryocut sections of fresh frozen tissues.

(2) Fix the sections in absolute or 80% alcohol for 10 minutes or in neutral formalin for 15 minutes.

(3) Treat the sections in 1N hydrochloric acid for 5 minutes.

(4) Rinse in distilled water.

(5) Treat the sections in 0.2M acetate buffer, pH 4.2 for 5-10 minutes.

(6) Transfer the sections in staining solutions and keep in a refrigerator overnight at about -4°C.

(7) Immerse in 0.05M acetate buffer (pH 4.2) for 10 minutes to remove unnecessary adsorption of the dye.

(8) Subject the sections to a second treatment with 0.05M acetate buffer, pH 4.2 in a bath for 10 minutes

(The stained sections can be left in dilute buffer, 0.05M acetate buffer (pH 4.2) for prolonged periods without loss of DNA bound dye).

(9) For permanent preparation, blot dry the sections and immerse them in absolute alcohol, clear in xylol and mount in Canada balsam or DPX.

Result: DNA sites will stain green or green-blue.

B. Removal of DNA from control sections by the use of deoxyribonuclease.

Theory: As a control for specific action of deoxyribonuclease, a serial section could be treated with the same incubation medium, for the same time, but without addition of deoxyribonuclease to the solution. A more rigorous control is to use the same incubation medium, containing deoxyribonuclease, but also containing zinc to inhibit the action of the enzyme, in an inactive form. Hence, DNA test is positive in the sections also.

Solution required

Incubation medium: The concentration of deoxyribonuclease should be 0.1%. Prepare 0.05M solution of cysteine hydrochloride (or hydroxylamine hydrochloride) in 0.05M sodium acetate solution at pH 7. Dissolve enough of deoxyribonuclease in this so that the final concentration of enzyme approximates to 0.1% (*e.g.* dissolve 2mg of enzyme in 1ml of cysteine-acetate buffer solution).

Adjust the pH. Leave the solution to stand at room temperature (or at 37°C) for 10-15 minutes. This will enable to inactivate the proteolytic activity.

Before use, dilute the solution with an equal volume of veronal-HCl buffer at pH 7 and to this add 1% of magnesium chloride.

Procedure

Treat the sections meant for the control with incubation solution at room temperature for an hour or so to ensure the complete removal of DNA from unfixed section and then proceed for staining with methyl green, as in case of test sections above.

Result: DNA positive sites of the test sections will remain unstained in control sections because DNA has been removed by deoxyribonuclease treatment.

E. HISTOCHEMISTRY OF NEUROHORMONES

Experiment No. 5

Histochemical demonstration of neurosecretary cells by Gomori's aldehyde fuschin staining technique as modified by Cameran and Stelle (1959).

Solutions and reagents

(a) Preparation of paraldehyde fuchsin stain

Add 1gram of basic fuchsin to 200ml of boiling water in a beaker. The boiling should not be continued for more than 1minute. Filter it and add 2ml of each of concentrated HCl and paraldehyde to the filtrate. Stopper the bottle, leave it at room temperature. Each day after a gap of 24hours withdraw just one drop from the bottle and place it on a filter paper. Fuchsin will decrease and the quantity of precipitate formed will go on increasing. When the solution has lost its reddish fuchsin colour in 3-4 filtrate. Allow the precipitate to dry on filter paper at the temperature of 60°C. Remove the crystalline precipitate from the filter paper and store it in a reagent bottle. The total yield will be roughly 1.9gram.

(b) Preparation of staining solution

Dissolve 0.25g of dry stain in 50ml of 70% alcohol. This solution can be stored for over 6months with no apparent change in its appearance and staining behaviour.

(c) Oxidant solution

$KMnO_4$	0.3gm
Distilled water	100ml
Conc. H_2SO_4	0.3ml

(d) Bleaching solution

Sodium bisulphate	4gm
Distilled water	100ml

Procedure

(1) Dissect out the nerve ring from a freshly narcotized earthworm and place it in a watch glass containing normal salt solution (0.73% NaCl).

(2) Fix it in aqueous Bouin's fluid for 12-24hours at room temperature.

(3) Wash well with distilled water.

(4) Treat with 70% alcohol and via descending grades of alcohol bring it back to distilled water.

(5) Treat the nerve ring with an oxidant (0.3g $KMnO_4$ dissolved in 100ml of water containing 0.3ml of H_2SO_4) for 2-3 minutes.

(6) Dry the nerve ring using blotting paper to wipe off the oxidant and bleach with 4% sodium bisulphite solution for 1-10 minutes or till the nerve ring is rendered translucent or white.

(7) Wash with distilled water for 5-10 minutes.

(8) Bring the nerve ring into 70% alcohol via 30% and 50% alcohol.

(9) Treat with staining solution for 2 minutes or more.

(10) Quickly blot off the excess of stain by means of filter paper.

(11) Differentiate with 90% alcohol until no more superfluous stain is given off. Change the alcohol twice or thrice. Even if after this much treatment any precipitate adheres to the ring then wash with 70% alcohol till the precipitate fades.

(12) Dehydrate in absolute alcohol, clear in xylol for 1-2hours or more till the material becomes semitransparent and sinks at the bottom of clearing fluid. Finally mount in Canada balsam or DPX.

Result: The neurosecretary cells will be rendered deep pink or violet and shall be distinctly visible against colourless background (Figure 3.2).

AUTORADIOGRAPHY OF RADIOACTIVE ELEMENTS

This is a technique to detect histochemically the presence of radioactive elements at the tissue or cellular level. It is a technique by which the localization of elements is ascertained provided they are: (a) radioactive, and (b) maintained at their original site during the preparation of the tissue and procedure for marking their location. (All depending upon the skill and intelligence of the operator).

The first requirement is fulfilled by treating the living organisms with radioactive isotopes; consequently if they are injected into or otherwise ingested by an animal, they become dispersed throughout the organism, mingled with non-radioactive elements. Once the radioisotopes have been incorporated into a tissue, their presence even in very low concentrations, can be detected by virtue of radiations which they emit.

In histochemistry this radiation can be detected and its source inside the tissue can be discovered by the fact that it affects photographic emulsions.

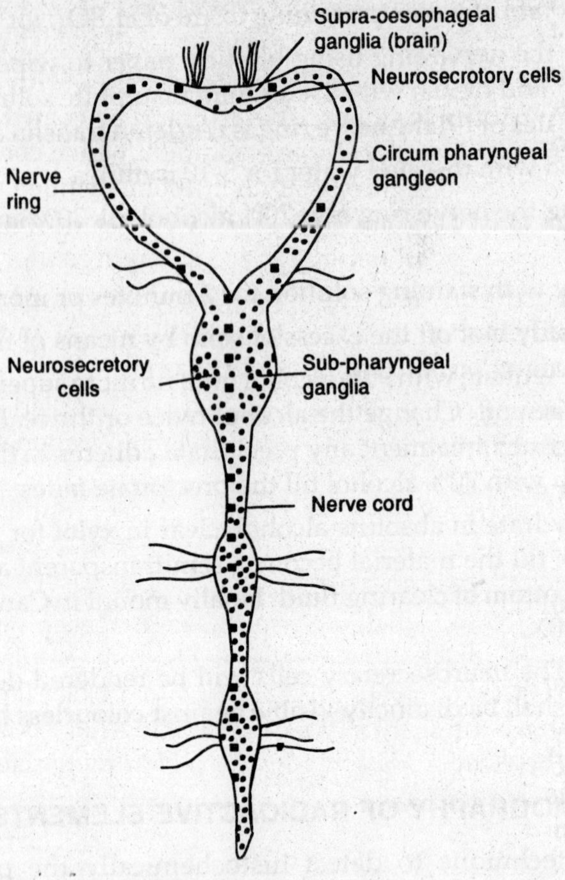

Fig. 3.2. Neurosecretory cells in the nerve ganglia of earthworm

Autoradiographic Methods and Uses

The autoradiographic methods are so valuable and so widely used that a wide range of techniques have been developed.

The most simple and widely used technique is: ***Stripping film technique.*** The procedure is detailed below.

(1) For studies dealing with DNA or proteins that can be fixed in the substances of the cell, fixed tissues are embedded in paraffin and sectioned normally, using routine methods. There are certain metals, used in some fixatives, which themselves react with photographic emulsions. Consequently, acetic, alcohol or formaldehyde–calcium are usually the recommended fixatives, prepared in the ratio of 1:3.

Cryostat sections of unfixed tissues can be used, but they should be fixed in acetic alcohol (For water soluble material it is essential to use unfixed cryostat sections).

(2) Hot water washed slides (not washed with acids) are used for picking sections. These should be coated (or subbed) by being dipped briefly in a solution, which contains 0.5g gelatin and 0.05g chrome alum in 100ml distilled water. The slide must be dried for 2days before it is used. These can be stored for prolonged periods at 4°C in a refrigerator.

(3) If paraffin sections are used, they are caused to adhere to the coated slide and then treated with xylol and taken through degraded concentrations of alcohol.

(4) The stripping film is prepared and handled in a dark room, with a little light as possible from a Wraften series No. 1, red safelight. First the edges of a photographic stripping – film plate are trimmed by tracing a cut all around the edge. Other cuts are made as shown in Figure 3.3a to produce eight segments of film.

(5) With the scalpel, pick up the edge of each segment of a film and very slowly strip that segment of the plate. Hold it with as little contact as possible and float it face (emulsion) downwards, on to the surface of water in a large bowl (Figure 3.3b). The temperature of the water should be maintained at 25°C to permit maximum expansion of film ($2\,^1/_2$ inch).

(6) Place the slide, section upper most, under the water and slowly raise it under the film (Figure 3.3c).

(7) Hang the slide (wrapped in film) pegged at one end, on a rock to drain and dry.

(8) When dry store in the light tight and dry container at 4°C. The film is now exposed to the radiation. The time of exposure can vary greatly, depending upon the radioactive substance used and the concentration, which it had achieved in the tissue. This duration may be for some days or some weeks.

(9) The film is developed on the section, all the process being done at 17-18°C to avoid damaging the film. It is essential to prepare your own D-19b Developer; Johnson's 'Fix-sol' in the ratio 1:9, which is most satisfactory for photographic fixations. [D19Developer (Analar Reagents): Elon - 2.2g; anhydrous sodium sulphate - 72g; Hydroquinone - 8g; anhydrous sodium bicarbonate - 48g; potassium bromide - 4g; distilled water - 1000ml. This should be filtered before use].

(10) The section can then be stained with haemotoxylin and eosin or with methyl green – pyronin. This should be done at 17-18°C.

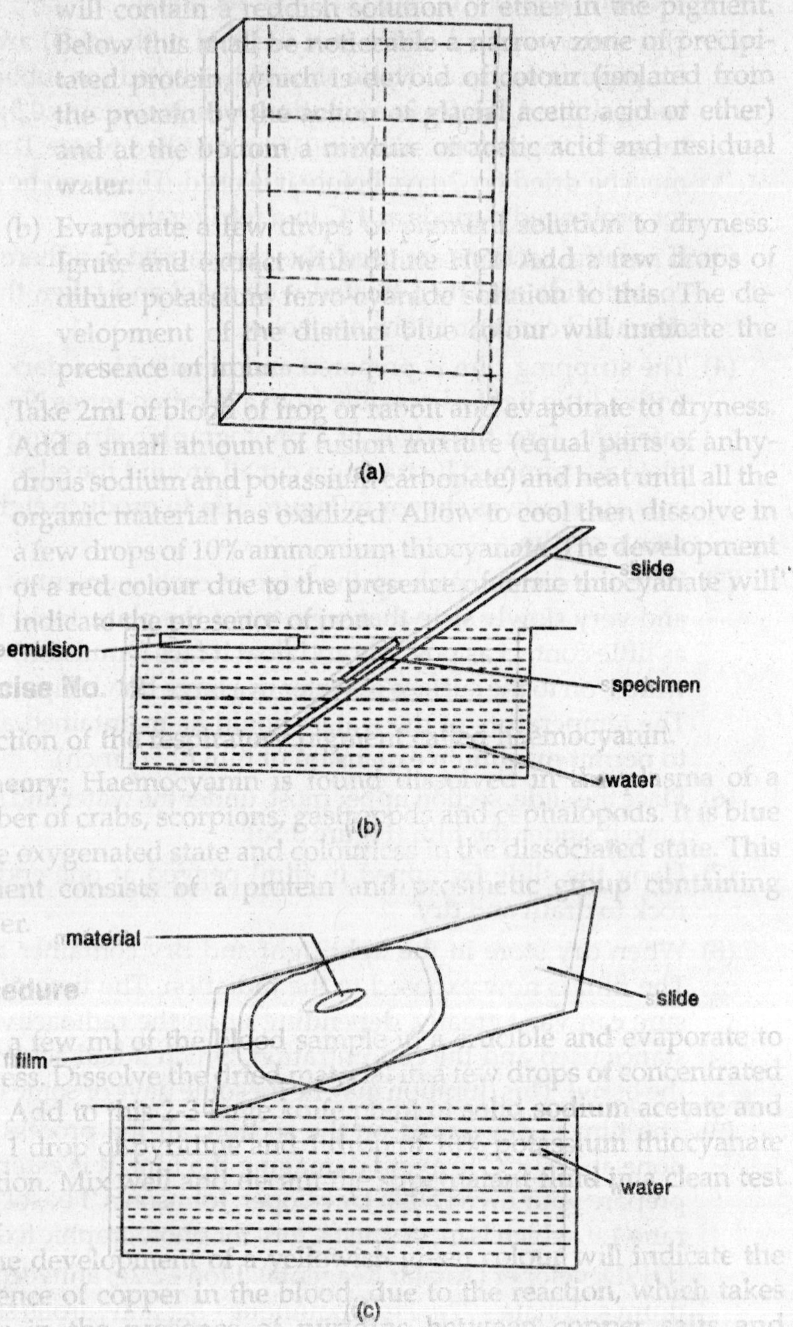

Fig. 3.3. Stepwise illustration of autoradiographic procedure

Results

The photographic emulsion acts as a coverslip for the sections but contains black regions or spots, caused by the radioactivity in the tissue sections below it. Hence, at one focus, the tissue histology can be inspected, while at a slightly higher focus you should be able to observe how much radioactivity has been emitted by each tissue or cellular component. Quantitatively, the number of photographic grains in the emulsion bears a definite relationship to the number of radioactive atoms in the tissue below that point in the emulsion.

4

Physiology of Enzymes

Any substance, which alters the speed of a reaction and remains unchanged at the end, is called a catalyst. The catalysts, which are produced by living organisms, are named as enzymes. All enzymes contain proteins hence they are most often referred to as protein catalysts.

Nothing very definite can be said about how enzymes act? However, it is believed that they attach themselves to the substrates and render these unstable. As the substrate breaks down, the enzyme involved is liberated and again attaches itself to more substrate molecules and the process is repeated. Thus, a small amount of enzyme reacts with a large amount of the substrate. The colloidal nature of the enzyme makes available large surface areas for the adsorption of the substrate.

Most of the enzymes are highly specific in their action. Some of them, such as *urease*, attack a specific compound and not the other. The others attack many different molecules having similar chemical structures, *e.g.* maltase, hydrolyase, α-glucosidase and emulsin β-glucosidase. Still others attack large groups of similar compounds, *e.g.* pepsin hydrolyse, all soluble native proteins.

Emil Fischer has explained the specificity by assuming that enzyme must fit into the substrate molecule as a key fits into a lock. Some enzymes are like specific keys, which fit into only one lock, whereas other enzymes are master keys, which fit into many locks.

The enzymes are classified into three main classes: the hydrolases, the oxidases and the desmolases. The hydrolases break up complex molecules into simpler ones by addition of molecules of water. The oxidases catalyze biological oxidations and reductions; and the desmolases break down C-C, C-N, and other linkages in the substrate molecules.

Here, the detection of enzymes has been based on two different methodologies.

I. Using Biochemical Methods

II. Using Histochemical Methods

I. BIOCHEMICAL METHODS FOR DETECTION OF ENZYMES

Experiment No. 1

Biochemical detection of the presence of *invertase* in tissue extract of the salivary glands of cockroach using Fehling's reagent.

Theory: This test is based on the fact that the sucrose used as a substrate is hydrolyzed by the enzyme *invertase* into glucose and fructose, which give positive tests with Fehling's solution. This is because the copper salts are reduced to cuprous oxide by sugars, such as fructose or glucose in alkaline medium. It is for this reason that sugars containing a free aldehyde or ketone group on boiling with Fehling's solution produce a characteristic brick red or brown precipitate of Cu_2O.

Solutions required

(a) Fehling's solution A

(b) Fehling's solution B

(c) 10% sucrose solution

Procedure

(1) For preparing the extract of the salivary glands of cockroach, take a few freshly anesthetized cockroaches, dissect them and take out their salivary glands. Rinse the salivary glands in distilled water and place them in a watch glass containing distilled water. An easy method of taking out large number of salivary glands is to pull the head of narcotized cockroaches along with alimentary canal and then to isolate and remove the salivary glands but hundred per cent success is not ensured by this method.

(2) Transfer the salivary glands in a small test tube containing 1-2ml of distilled water and grind them as completely as possi-

ble. Allow the test tube to stand for a while and then remove the clear supernatant fluids in another test tube by decantation process.

(3) Now, take three test tubes and label them as A, B and C.

(4) In each of the test tube, take 2ml of 10% sucrose solution.

(5) Now proceed as under:

In test tube A, add 3drops of salivary gland extract. In test tube B, add 3drops of salivary gland extract that has already been boiled (by placing a test tube containing extract in boiling water in a beaker for 10-15 minutes with a view to inactivate the enzyme). In test tube C, nothing more to be added.

(6) Place all the test tubes in a beaker and place the beaker itself in an oven for incubation at 37°C for a period of 24hrs or simply overnight.

(7) Now, the contents of each test tube A, B, C should be separately tested by adding one drop of Fehling's solution A and one drop of Fehling's solution B and again putting them together in a beaker containing boiling water or water bath and observe the colour changes as under:

Observation

Test tube A. A brown or brick red precipitate is formed

Test tube B. No change in colour or contents

Test tube C. No change in colour or contents

Inference and Results: In test tube A, the development of a brick red or brown precipitate (a positive Fehling's Test) indicates the presence of glucose formed due to hydrolysis of sucrose by enzyme invertase, whereas in test tube B, no change occurs due to the presence of inactive enzyme and in test tube C, no change occurs due to total absence of the extract or enzyme.

Experiment No. 2

Biochemical detection of invertase in the extract of the salivary glands of cockroach using Benedict's solution.

Solutions required

Benedicts solution

Copper sulphate	17.3g
Sodium citrate	173.0g

Sodium Carbonate anhydrous 100.0g

(a) Dissolve 173.0g of sodium citrate and 100.0g sodium carbonate in 800ml of distilled water in a beaker by heating.

(b) Dissolve 17.3g of copper sulphate in 100ml of distilled water. Add this to (a) and make up the volume to 1.0litre in a volumetric flask.

Procedure

Proceed as in experiment No. 1 upto step 6 and thereafter as under from step 7 onwards.

(7) Now, test the contents of test tube A, B, and C separately by using Benedict's solution. Take another set of test tubes mark them as A_1, B_1, and C_1, and in each test tube take 5ml of Benedict's solution and add 7 - 8drops from test tubes A, B and C into test tubes A_1, B_1, and C_1, respectively. Mix well and bring the contents of each test tube to boiling for two minutes or for 3-4 minutes by placing the test tubes A_1, B_1, and C_1 in boiling water in a beaker.

(8) Take out the test tubes and allow them to stand.

Observation

In test tube A_1, the precipitate formed will be red, yellow or green.

In test tube B_1, there will be no change.

In test tube C_1, there will be no change.

Inference and result: The formation of red, yellow or green precipitate in test tube A_1, will indicate the presence of glucose, which has been produced by hydrolysis of sucrose due to the presence of enzyme invertase in the salivary gland extract of cockroach. In test B_1, no change occurs due to inactive enzyme and in test tube C_1, no change occurs due to absence of enzyme

Experiment No. 3

Biochemical detection of amylase in the extract of the salivary glands of cockroach or in your own saliva using starch as substrate and iodine solution as colouring agent.

Solutions required

(a) ***1% Starch solution***

Starch 1g

Distilled water	100ml

(b) ***Lugol's iodine solution***

Iodine	1g
Potassium iodide	2g
Distilled water	100ml

Procedure No. 1: *For the test of amylase from the extract of salivary glands of cockroach.*

(1) Prepare the extract of salivary glands of cockroaches as in step No. 1 of experiment No. 1

(2) Take three test tubes and mark them as A, B, and C. Place 1ml of 1% starch solution in each and then proceed ahead as under.

(3) In test tube A, add 10drops of extract and 1drop of 1%NaCl. Mix well.

(4) In test tube B, add 10drops of extract and 1drop of 1%NaCl and boil the mixture to kill or inactivate the enzyme.

(5) In test tube C, add 10drops of distilled water and 1drop of 1%NaCl.

(6) Place all the test tubes in a hot air oven for incubation at 37°C for 24 hours.

(7) Now, add 1-2drops of Lugol's iodine in each test tube A, B, and C and shake it gently. Observe the development of colour or change in each.

Observation

In test tube A, the solution did not turn blue showing the absence of starch in A, as it was hydrolyzed to glucose by the action of enzyme amylase.

On the other hand, in test tube B, the contents will become blue as the starch remained unaffected due to inactivation of enzyme and in test tube C, also the contents will become blue because here also the starch remained unaffected due to the total absence of enzyme as no extract was added to it.

Inference and Result: In test tube A, blue colour did not develop as starch was hydrolyzed to glucose by amylase, while in B, C blue colour developed as starch remained unaffected due to inactivation of amylase in B or its total absence in C.

The observation and inference drawn above can be confirmed by testing the contents of test tubes A, B and C after step No. 4 by testing with Fehling's solution or Benedict's solution in which case only positive test of these two shall be available with contents of test tube A and negative with contents of test tube B and C.

Procedure No. 2: *For the test of amylase from the saliva of human beings.*

Collection of saliva: For collecting your own saliva either keep the mouth closed for a while, resulting in accumulation of some saliva in your buccal cavity or keep your mouth closed and chew two small pieces of a lemon, a lot of saliva will collect inside the buccal cavity or chew a few pieces of paraffin wax and after a few minutes collect the saliva in a clean beaker. Dilute the saliva with an equal volume of distilled water and filter it using cotton, if required. The filtrate should be used as a source of enzyme amylase for testing.

(1) Take three test tubes and label them as A, B and C.
(2) In each test tube place 5ml of 0.5% or 1% freshly prepared starch solution and proceed ahead as under.
(3) In test tube A, add nothing.
(4) In test tube B, add 5ml of dilute saliva.
(5) In test tube C, add 5ml of dilute saliva, which has already been boiled to inactivate the enzyme.
(6) Place all the three test tubes in a hot air oven for incubation at 37°C for 24 hours or less.
(7) Take another set of three test tubes and mark them as A, B and C and in each place 1-2 drops of Lugol's iodine and thereafter in each tube add 8-10drops from the respective tubes A, B, and C. Mix the contents of each well and allow to stand and cool down and then observe the changes in the contents of each.

Observation

In test tube A, there will be no change in colour but on addition of a few drops of Lugol's iodine, the colour will become blue indicating the presence of starch with no change.

In test tube B, no colour will be developed on addition of Lugol's iodine, because due to the action of enzyme, amylase, starch has been converted into glucose.

In test tube C, again there will be no change due to inactivation of amylase but a blue colour will develop on addition of a few drops of Lugol's iodine indicating the unchanged state of starch.

Result: The absence of blue colour in test tube B indicates that amylase of saliva has hydrolyzed the starch to glucose, hence negative to Lugol's iodine test.

Experiment No. 4

Biochemical detection of protease (pepsin and trypsin) from the animal source or commercial preparations using Biuret Test.

Theory: For experiments, either a peptic extract of the mucous membrane of a dog or pigs stomach or for laboratory experiments frog's stomach or commercial pepsin is used. Some workers have even used the extracts of the different parts of the alimentary canal of cockroach. The present concept is that pepsin in the digestive tract hydrolyses proteins to proteoses and peptones.

Procedure

Procurement of pepsin from cockroach

Dissect out the midgut of a large number of freshly anesthetized cockroaches, rinse in distilled water, slit open longitudinally, remove the contents and grind it thoroughly well and dilute the grinded paste with 2.0ml of distilled water or use the commercial pepsin (which is available in the form of a crude extract of the gastric mucosa of cattle).

(1) Take five test tubes and label them as A, B, C, D, and E.

(2) In each test tube, add a small quantity or portion of the white of half boiled egg or threads of fibre.

(3) Now proceed ahead as under:

   (i) In test tube A, pour 5ml of 0.75% aqueous solution of pepsin and mix it with 5ml of 0.5% HCl.

   (ii) In test tube B, pour 5ml of 0.75% aqueous solution of pepsin and 5ml of distilled water.

   (iii) In test tube C, pour 5ml of 0.5% HCl and 5ml of distilled water.

   (iv) In test tube D, pour 5ml of 0.75% aqueous pepsin solution, which has already been boiled to inactivate the enzyme and 5ml of 0.5% HCl.

   (v) In test tube E, pour 5ml of 0.75% aqueous solution of pepsin and 5ml of 40% NaOH.

(4) Place all these labelled test tubes in a beaker, which has to be placed in a water bath set at 37°C to 40°C for 45-60 minutes.

(5) Now, remove one by one each test tube and test a portion of the contents of each by applying Biuret test (see foot note) without boiling and observe the following changes.

Observation

(i) The contents of test tube A are rendered rose pink in colour indicating the conversion of protein into peptones.

(ii) The contents of test tube B, no change because hydrolysis of protein could not take place due to absence of HCl.

(iii) The contents of test tube C, no change because of absence of enzyme pepsin, hence no hydrolysis of proteins to peptones.

(iv) The contents of test tube D, no change because boiling, hence no hydrolysis and no colour development inactivated the pepsin.

(v) The contents of test tube E, no change because HCl was absent, which was essential for the hydrolytic action of pepsin.

Inference and Results: The proteins are hydrolyzed to peptones by pepsin only in acidic medium.

For testing the contents of each test tube by Biuret test, take separately 2ml of the solution from each test tube and mix with it 2ml of 40% NaOH solution. Thereafter, add to it one drop of 1% copper sulphate solution. The development of pink or purple colour will indicate the presence of peptones, the hydrolytic products of protein of egg albumin.

Experiment No. 5

Biochemical detection of trypsin in the extract of pancreas of pig using Biuret test.

Theory: Pancreatic juice contains trypsin, a proteolytic enzyme; (the commercial preparation of trypsin is called ***pancreatin***). A simple source of enzyme trypsin is to prepare an extract of the pancreas of pig by grinding it in a little quantity of distilled water and then diluting it.

Procedure

As a source of trypsin either use 5% aqueous solution of ***pancreatin*** or trypsin or diluted solution of the extract of pancreas.

(1) Take three test tubes and label them as A, B, and C,

(2) In each test tube pour 5ml of pancreatin solution, and a small piece of the egg white of a boiled egg or simply the fibres and proceed ahead as under.

(3) In test tube A, add 5ml of 1% sodium carbonate solution.

(4) In test tube B, add 5ml of distilled water.

(5) In test tube C, add 5ml of 0.5% HCl.

(6) Put all the test tubes in a beaker and place the beaker for incubation at 37-40°C in a water bath for 45-60 minutes.

(7) Test separately the contents of each test tube for the presence of peptones by applying the Biuret test as mentioned earlier and observe the changes as under.

Observation

In test tube A, there is no change

In test tube B, there is no change

In test tube C, the contents are turned pink to purple indicating the presence of peptones due to hydrolysis of proteins by protease, trypsin.

Experiment No. 6

Biochemical detection of proteases in the midgut extract of cockroach using capillary tubing method.

Procedure

(1) Dissect out the midgut of a freshly narcotized cockroach and after cleaning grind it in a pastle-mortor with a little quantity of water and dilute it with 20ml of distilled water.

(2) Fill up a few capillary tubings with the egg albumin and dip them in warm water in a beaker. The albumin will coagulate, (while dipping the capillary tubings in egg albumin care should be taken to avoid the entry of air bubbles). Break the capillary containing coagulated albumin into five pieces.

(3) Take two test tubes and label them as A and B.

(4) In test tube A, pour 1ml of phosphate buffer (pH 8.04) solution and add 1ml of unboiled diluted extract of midgut. In test tube B, pour 1ml of boiled extract of midgut.

(5) Put two pieces of capillary tubings filled with coagulated albumin in each test tube A and B.

(6) Add 10drops of toluene in each test tube.

(7) Put both the test tubes in a beaker, which should be placed inside an incubator at 37°C for 24 hours.

Observation and Inference

Hold the test tube A and B separately against the light and notice that the capillary tubings in test tube A have become almost transparent showing the digestion of albumin by protease, while that of test tube B are opaque indicating the presence of undigested albumin because in this test tube the midgut extract was boiled to inactivate the enzyme.

Experiment No. 7

Biochemical detection of lipases in midgut extract of cockroach using phenol red or toluene.

Procedure 1

(1) Take three test tubes and label them as A, B and C.

(2) In test tube A, put 5ml of unboiled extract (as prepared in previous exercise).
In test tube B, put 5ml of boiled extract.

(3) In test tube C, simply put 5ml of distilled water.

(4) Now, in test tube A and B only, put boiled milk and phenol red separately. Thereafter, add drop-by-drop 0.025% Na_2CO_3 solution in each tube A and B till it becomes pink. Incubate at 37°C till the contents of test tube A become colourless.

Observation and Inference

The contents of test tube A are rendered colourless because the milk fat has been hydrolyzed by the enzyme lipase of extract whereas in B the pink colour persisted because enzyme of extract was destroyed by boiling.

Procedure 2

(1) Dilute the condensed milk with distilled water so as to obtain its 10% solution.

(2) Take 5ml of 10% solution of condensed milk and add 3-4drops of bromo thymol blue in it. Thereafter pour 1% NaHCO3 gradually until turns light blue. In the test tubes A and B,

which contain unboiled and boiled tissue extracts, add blue milk solution to each and few drops of toluene.

(3) Incubate for 3-4 hours. The blue colour of the contents of test tube A will change to yellow through green due to digestion of milk fat by the enzyme, lipase.

Observation and Inference

The development and change of colour in test tube A is due to hydrolysis of the milk fat by lipase of extract. The absence of colour development in test tube B is due to inactivation of enzyme in extract.

II. HISTOCHEMICAL METHODS FOR DETECTION OF ENZYMES

(A) Alkaline Phosphatase

Experiment No. 1

Histochemical detection of alkaline phosphatase in the frozen dried sections or unfixed cryocut sections of kidney or liver of vertebrates using the Gomori-Takamatsu procedure.

Solutions required

(a) Incubation medium

2% barbitone sodium (sodium 5, 5-diethyl barbiturate)	10ml
2% sodium β-glycophosphate	10ml
2% Calcium chloride dried	20ml
(or 2.7% $CaCl_2$. 2H2O)	
5% magnesium sulphate ($MgSO_4.7H_2O$)	1ml
Distilled water	5ml
pH of this medium should be 9.4	

(b) Cobalt-nitrate solution

Cobalt nitrate	2gm
Distilled water	100ml

(c) Yellow ammonium sulphide solution

Yellow ammonium sulphide	0.5ml
Distilled water	95.5ml

Procedure

(1) Prepare cryocut sections (8-10μ) of fresh frozen tissue (kidney or liver).

(2) Transfer the sections in the incubation medium contained in a cavity block and incubate the sections at 37°C for upto 60 minutes (the incubation period of kidney varies from 10-15 minutes, while that of liver may require 60 minutes).

(3) Remove the incubation medium and wash the sections in running tap water for 5 minutes.

(4) Immerse the sections in 2% cobalt nitrate solution for 5 minutes.

(5) Rinse the sections in distilled water.

(6) Immediately immerse the sections in 0.5%(v/v) yellow ammonium sulphide solution just for 1minute.

(7) Wash the sections in distilled water.

(8) Mount in glycerine jelly.

Result: The enzyme activity is shown by a black precipitate of cobalt sulphide particularly in brush borders of kidney tubules (Figure 4.1).

For confirmation of the results, a parallel set of control sections should be run along with test sections. In control sections, 10ml of 3% sodium β-glycophosphate in incubation medium should be replaced simply by 10ml of distilled water.

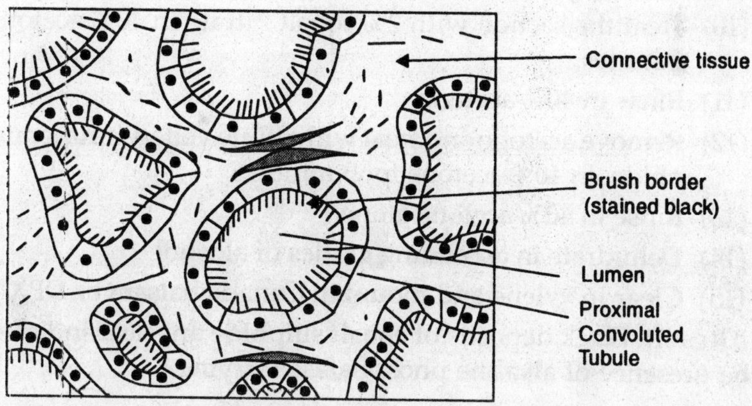

Fig. 4.1. T.S. of kidney of rat

Experiment No. 2

Histochemical detection of alkaline phosphatase in the paraffin sections of kidney or liver of vertebrates using Gomori method as modified by Fredricsson (1952 & 1956).

Procedure

(1) Fix small pieces of tissue in 90% ethanol for 24 hours at 22°C.
(2) Treat the fixed tissue twice with absolute alcohol, each time for 40-60 minutes.
(3) Treat the tissue twice with benzene each time for 30 minutes.
(4) Transfer the tissue in embedding pan containing molten wax (m.p. 56°C) for infiltration of wax inside the tissue. At least the molten paraffin wax should be changed three times, each time for 20 minutes. Total embedding period should in no case be less than 60 minutes.
(5) Prepare the paraffin blocks of the tissue and then trim the blocks to proper dimensions.
(6) Prepare microtome sections (8-10μ) and directly mount on the slide without using water. Ensure that sections have taken grip on the surface of slide.
(7) De-wax the section with xylene and bring them to water via absolute alcohol and 40% acetone.
(8) Place the slides in coupling jar containing incubation medium and put in a temperature bath for 20-60 minutes at 37°C. Keep the coupling jar closed by the lid.
(9) Rinse quickly in 40% acetone.
(10) Treat the section with 2% cobalt nitrate in 40% acetone for 5 minutes.
(11) Rinse in 40% acetone.
(12) Remove acetone and treat with dilute yellow ammonium sulphide in 40% acetone for 3 minutes.
(13) Rinse in 40% acetone alone.
(14) Dehydrate in ascending grades of alcohol
(15) Clear in xylene and mount in Canada balsam or DPX.

Result: Black deposits of cobalt sulphide are clear indications of the presence of alkaline phosphatase enzyme.

Experiment No. 3

Histochemical detection of alkaline phosphatase in the paraffin sections of kidney or liver of vertebrates using Gomori Calcium method (Pearse, 1968).

Solutions required

All as in experiment No. 1.

Procedure

(1) Fix small pieces of tissue in three changes of cold acetone for 24-48 hours from freshly narcotized animals.

(2) Dehydrate the tissue in ascending grades of alcohol, for 30 minutes in each grade.

(3) Transfer the tissue in absolute alcohol (1:1, v/v) for 30 minutes.

(4) Clear in xylene or benzene and transfer in the embedding pan containing molten wax (m.p. 52°C or 56°C) for at least 8-12 hours.

(5) Prepare the paraffin blocks and after proper trimming cut sections of 8-10μ thickness.

(6) Flatten the sections, dry the slides for 3 hours at 37°C and store till further use or proceed ahead as under.

(7) Treat the sections with petroleum or chloroform for 1minute.

(8) Bring the sections in distilled water by subsequent treatment involving 3-4 dips in each absolute alcohol, acetone, 80% and 50% acetone - water, rinse with several dips in distilled water.

(9) Treat the sections with incubation medium at 37°C for 16 hours.

(10) Thoroughly wash with distilled water.

(11) Treat the section with 2% cobalt nitrate or cobalt acetate for 3-4 minutes.

(12) Rinse in distilled water.

(13) Treat the sections with diluted 1% yellow ammonium sulphide solution for 1-2 minutes (Avoid prolonged treatment).

(14) Wash in distilled water.

(15) Dehydrate in ascending grades of alcohol.

(16) Clear in xylol and mount in Canada balsam.

Result: The sites showing alkaline phosphatase activity are stained brownish black.

Experiment No. 4

Histochemical detection of alkaline phosphatase in fresh frozen cryocut sections of kidney of vertebrates using the Naphthol phosphate method (Gomori, 1952).

Solution required

Incubation medium

2% barbitone-sodium (sodium 5, 5-diethyl barbiturate)	25ml
α-naphthol acid phosphate (sodium salt)	10mg
10% magnesium chloride	0.2ml
Fast blue RR salt (Stir and filter into a coupling jar and use immediately)	25mg

The pH of required incubation medium is 9.2

Procedure

(1) Prepare fresh frozen cryocut sections of the tissue (preferably kidney of a freshly narcotized animal).

(2) Transfer the sections in a cavity block and incubate them in the incubation medium for 15-20 minutes at 37°C.

(3) Wash well with distilled water.

(4) Wash the section in 1% acetic acid solution for 1minute.

(5) Rinse well in distilled water.

(6) Mount in glycerine jelly.

Run a set of control section with test section but without α-naphthol acid phosphate in the incubation medium.

Results: The sites of alkaline phosphatase activity are indicated by the deposition of a purple-black precipitate.

(B) Adenosine Triphosphatase

Theory: This enzyme is also referred to as the *'calcium-activated ATPase'* rather than as *'myosin ATPase'*. Although it is an inherent attribute of myosin molecule, it occurs in liver; kidney (very high, concentrated); heart (very high, concentrated); skeletal (high, concentrated); lung (high, concentrated); pancreas and brain of rat.

Experiment No. 1

Histochemical detection of acid triphosphatase in fresh unfixed cryocut sections of animal tissue using calcium method.

Solution required

Incubation medium

0.1M Sodium barbiurate solution (2.06g/100ml distilled water)	10ml
ATP	76mg
2,4-dinitrophenol	30mg
calcium chloride solution (10.5g $CaCl_2.2H_2O$/ 100ml of distilled water)	5ml
Distilled water	35ml

(Prepare immediately before use)

Procedure

(1) Prepare fresh cryocut section of the tissue from a freshly narcotized or slaughtered animal.

(2) Transfer the sections in a cavity block containing the freshly prepared incubation medium for 10-30 minutes at 37°C.

(3) Wash the sections thrice with 1% calcium chloride solution, each time for 2 minutes. (Calcium ions keep the precipitate formed from dissolving).

(4) Remove calcium chloride solution and treat the sections with 1% cobalt nitrate solution for 1-2 minutes and thereafter again treat the sections twice with fresh 1% cobalt nitrate each time for 2 minutes (The total duration of this treatment should not exceed from 6-8 minutes). By this treatment the calcium phosphate is changed into cobalt phosphate.

(5) Wash the sections thoroughly well with distilled water to remove the adsorbed cobalt.

(6) Treat the sections with 0.5% (v/v) yellow ammonium sulphide for just a minute taking care that sections remain completely submerged in solution.

(7) Finally wash well in distilled water.

(8) Mount in glycerine jelly.

Control: A set of control sections should be run parallel to test sections. In sections of muscle, it is simple to control this reaction by incubating a serial section under the identical conditions but with β-glycerophosphate instead of the adenosine triphosphate.

Results: The sites of ATPase activity are revealed by the deposition of a brown-black precipitate of cobalt sulphide.

Experiment No. 2

Histochemical detection of acid phosphatase in fresh frozen cryocut sections of animal tissue using lead method (Watchstein and Meisel, 1956; Watchstein et al., 1960).

Solution required

Incubation medium

ATP	25mg in 20ml distilled water
O.2M trisbufffer pH 7.2	20ml
2% lead nitrate	3ml
0.25% magnesium sulphate	5ml
Distilled water	2ml

Note: 40mg of dinitrophenol (DNP) can be added to this medium as it activates the activity of this enzyme.

Procedure

(1) Prepare fresh frozen cryocut sections (8μ-10μ) and transfer them in a cavity block.

(2) Treat the sections with incubation medium for 2 hours at 37°C.

(3) Wash sections in running tap water.

(4) Quickly rinse in distilled water.

(5) Immerse and submerge the sections in 0.5% (v/v) yellow ammonium sulphide solution for just 1minute.

(6) Wash thoroughly well in distilled water.

(7) Mount in glycerine jelly.

Results: ATP positive sites are indicated by the deposition of a brown precipitate of lead sulphide.

(C) LIPASE

Experiment

Histochemical detection of lipase in animal tissue using Tween method of Gomori (1952) as modified by Fairweather.

Theory: The substrate for lipase is a Tween. The enzyme lipase splits the fatty acids off and this is trapped as insoluble calcium salt. The latter is then converted into lead salt in order to produce a dark sulphide, which stains the tissue.

Solutions required

(a) Stock solutions

A. 5% aqueous solution of the selected Tween 40, 60 or 80 (Tween 80 gives the best results with true lipase)

B. 0.5M tris buffer, pH 7.2 or pH 7.4

C. 10% solution of calcium chloride (to be prepared afresh)

(b) Incubation medium

Tris buffer (B)	5ml
Tween solution (A)	2ml
Calcium chloride solution (C)	2ml
Distilled water	40ml

Filter this mixture and add a crystal of thymol to the filtrate for preservation and storage.

Procedure

(1) Prepare fresh cryocut sections (8μ-10μ) of the tissue of a freshly narcotized or slaughtered animal.

(2) Transfer the sections in a cavity block containing incubation medium at 37°C for 2-4 hours

(3) Wash the sections in distilled water, preferably containing 1% calcium chloride to ensure that calcium soaps remain insoluble.

(4) Submerge the sections in 1% solution of lead nitrate for 15 minutes at 37°C.

(5) Wash in running tap water for 5 minutes.

(6) Immerse the sections for 1-2 minutes in water saturated with hydrogen sulphide (or 1% yellow ammonium sulphide).

(7) Rinse in distilled water.

(8) Wash in running tap water

(9) Mount in glycerine jelly otherwise for permanent preparation dehydrate in ascending grades of alcohol, clear in xylene and mount in Canada balsam or D P X.

Control: Run a set of control sections parallel to test sections but the incubation medium should not contain Tween.

Results: The sites showing the deposition of brown-black precipitate of lead sulphide indicate the presence of lipase activity.

(D) PROTEASES

Theory: The proteases are also referred to as peptidases, which are capable of hydrolysing protein and amino acid compounds and carboxyl group. These peptidases are differentiated as exopeptidases and endopeptidases. However, phosphamidase (endopeptidase) and aminopeptidase (exopeptidase) reactions are supposed to pro vide more valuable informations for the whole range of proteolytic enzymes (Kidney or liver of rat).

Experiment No. 1

Histochemical detection of aminopeptidase in animal tissue using Chelation method (Nachlas *et al.*, 1957).

Solution required

Incubation medium

0.1M acetate buffer pH 6.5-6.8	10ml
0.85% solution of sodium chloride	8ml
0.02M (32mg in 25ml of water) of potassium cyanide	1ml
Leucyl-β-naphthyamide	8mg in 1.0ml of distilled water
Just before use add: Fast blue B	10mg

Procedure

(1) Prepare fresh cryocut sections (8μ-10μ).

(2) Treat the sections in a cavity block with incubation medium at 37°C for 15 minutes.

(3) Replace the incubation medium with fresh incubation medium for another 15 minutes. This process should be repeated until sufficient reactions.
(The sections of kidney respond strongly in 15 minutes, whereas those of liver require 2 hours for significant response).

(4) Rinse the sections with 0.85% solution of sodium chloride at room temperature.

(5) Transfer the sections to 0.1M solution of copper sulphate (2.5g of $CuSO_4.5H_2O$ or its equivalent in 100ml distilled water) for 2 minutes at room temperature.

(6) Rinse the sections with 0.85% solution of sodium chloride at room temperature.

(7) Mount in glycerine jelly.

Control: Although there is no adequate control but sections can be incubated in the absence of substrate or the incubation medium as such can be used in which the enzyme activity has been inhibited by treatment with absolute alcohol at 37°C for 30 minutes.

Results: Blue lysosomes like particles or purple stains indicate the presence of enzyme activity.

Experiment No. 2

Histochemical detection of aminopeptidase in animal tissue using coupling method (Burstone and Folk, 1956).

Solution required

Incubation medium

1% L-Leucyl-β-naphthyamine	1.0ml
distilled water	40.0ml
0.2M Tris buffer pH 7.1	10.0ml
Garnet GBC	30.0ml

This mixture should be prepared afresh and before use it should be filtered.

Procedure

(1) Prepare fresh frozen sections (8μ-10μ) of the kidney or liver of rat and transfer in a cavity block in water.

(2) Rinse the sections with distilled water.

(3) Incubate the sections at 37°C for 20-60 minutes.

(4) Wash with running tap water for 5 minutes.

(5) Mount in glycerine jelly.

Results: The sites of leucine aminopeptidase activity are rendered red.

(E) GLYCOSIDE HYDROLASES, ESPECIALLY β-GLUCURONIDASE

Theory: The hydrolases are especially important because many of them are associated with the digestion of foods in the body. The hydrolases are concerned with the breakdown of large food molecules into smaller ones by hydrolysis. The hydrolases also include the disaccharide splitting enzymes, particularly the glycoside hydrolases, which include β-glucuronidase, β-glyconidase, β-galactosidase, acetyl-β-glucosaminidase and disaccharidase, *etc.*, which are significant from the histochemical viewpoint.

Of all these, β-glucuronidase is of special histochemical interest because it occurs in most tissues of mammals and has been reported in abundance in liver, kidney, spleen, lung, adrenal, thyroid and uterus. Within cells its localization appears mainly in the mitochondrial (or lysosomal) and microsomal fractions. The hydrolytic activity of β-glucuronidase varies from pH 3.5-5.0, *i.e.*, their optimal pH is in the acid range.

The history of the histochemical data of β-glucuronidase indicates that the enzyme is readily liberated into the general cytoplasm, which becomes coagulated during fixation; the enzyme rich granules in such a section appeared to be denser aggregates of cytoplasmic-diffuse enzyme coagulum, with the enzyme also distributed generally throughout the cytoplasm.

Experiment No. 1

Histochemical detection of β-glucuronidase using simultaneous coupling method (Hayashi et al., 1964). This is also referred to as Naphthol AS-BI Method (Hayashi et *al.*, 1964).

Solutions required

(a) Substrate Stock solution

Dissolve 28mg of naphthol AS-BI glucuronide in 1.2ml of 0.05M-sodium bicarbonate (420mg sodium bicarbonate in 100ml of distilled water). Add 0.2M acetate buffer at pH 5.0 and make up the volume to 100ml. This solution is stable for several weeks at room temperature.

(b) Pararosaniline solution

Dissolve 1g of pararosaniline hydrochloride in acidified distilled water (prepared by mixing 5ml of concentrated HCl with 20ml of distilled water) slowly with gentle warming. Allow it to stand to cool down. Filter and store at room temperature.

(c) Sodium nitrate solution

Prepare 4% sodium nitrate solution (4.0g sodium nitrate in 100ml of distilled water). This should always be made afresh.

(d) Incubation medium

First add 0.3ml of pararosaniline solution B to 0.3ml of the sodium nitrate solution C in the beaker. Then after 1minute, add 10ml of substrate stock solution A in a 20ml beaker. Shake the mixture for a few seconds until the colour becomes amber.

Before use adjust pH to 5.2 by adding a few drops of caustic soda solution (N-NaOH). After this, make the volume of the final mixture upto 20ml with distilled water. Filter through a Whatman No. 1 filter paper.

Procedure

(1) Fix the slices of tissue (liver, kidney, adrenal, *etc.*) in formol-calcium (4%, w/v formaldehyde and 1% anhydrous calcium chloride; w/v) at 4°C for 24 hours.

(2) Do not wash, only blot on filter paper. Transfer to Holt's hypertonic gum-sucrose medium (0.88M sucrose containing 1% gum acacia) for at least 24 hours at 4°C.

(3) Cut frozen sections (8μ-10μ) and process them in free floating condition either on a slide inside a ring or in a small cavity block.

(4) Treat the sections with incubation medium at 37°C for 20-30 minutes (only in case of kidney and liver of rat).

(5) Rinse the sections in distilled water.

(6) Mount free floating sections on to slide.

(7) Dehydrate in alcohol, clear in xylol and mount in DPX or any other suitable mounting medium.

Results: The presence of predominantly red stained small granules indicate the sites of the enzyme activity.

Experiment No. 2

Histochemical detection of β-glucuronidase in animal tissue using Azo Dye method (Rath and Otto, 1966).

Solution required

Incubation medium

Dissolve 15mg of 8-hydroxyquinoline-b-glucuronide in 1ml of ethyl alcohol and add 4ml of 200mM acetate buffer (pH 5.2) to it. Shortly before use mix 20mg of Fast block salt K dissolved in 15ml of distilled water.

Procedure

(1) Prepare cryocut sections (8μ-10μ) of unfixed fresh frozen tissue (kidney and liver of rat) and transfer them in a cavity block containing distilled water.

(2) Rinse in distilled water and incubate the sections in incubation medium at 37°C for 2-6 hours.

(3) Wash the sections thoroughly with distilled water.

(4) Immerse the sections in 1% aqueous solution of sulphanulic acid for 8-10 minutes.

(5) Wash well in distilled water and mount in glycerine jelly.

Results: The sites showing the deposition of dark brown precipitate indicate positive β-glucuronidase activity.

Experiment No. 3

Histochemical detection of β-galactosidase in animal tissue (kidney or liver of rat) using Post-Coupling method (Rutenburg *et al.*, 1958).

Solution required

Incubation medium

6-bromo-2-naphthol B D galactopyranoside	100.0mg
Methyl alcohol	15.0ml
Hot distilled water	200.0ml
Mix in the above order, allow it to cool down and then add	
Phosphate citrate buffer (pH 4,95)	85.0ml

Prepare phosphate citrate buffer by mixing equal volumes of 0.2m disodium phosphate and 1.0M citric acid. Further add 100ml of distilled water. This can be stored for upto 4-6months at 4°C.

Procedure

(1) Prepare fresh frozen cryocut sections (8μ-10μ) of the kidney or liver of a freshly narcotized rat and transfer them in a cavity block containing distilled water.

(2) Incubate the sections at 37°C for 2-8 hours.
(3) Wash thrice with distilled water, each time for 3 minutes.
(4) Transfer the sections to a freshly prepared cold solution (4°C) of Fast blue B (1mg/ml) at pH 7.4-7.8 for 4-5 minutes with gentle agitation.
(5) Wash thrice with cold distilled water.
(6) Mount in glycerine jelly.

Results: The development of blue or purple colour indicate the presence of β-galactosidase activity.

Experiment No. 4

Histochemical detection of the Alpha-galactosidase in the liver or kidney of rat using Post-Coupling method (Rutenburg *et al.*, 1960).

Solution required

Incubation medium

6-bromo-2-naphthol-α-D-galactopyranoside	15.0mg
Ethyl alcohol (96%)	10.0ml
0.025M phosphate buffer pH 6.5	90.0ml

Procedure

(1) Prepare the slices of the tissue, fix them in cold neutral formaline for10-12 hours.
(2) Cut fresh frozen sections (8μ-10μ) and transfer them in a cavity block.
(3) Incubate the sections in free-floating condition at 37°C for 1-2 hours.
(4) Rinse in distilled water.
(5) Transfer the sections in 1mg/ml Fast blue B solution prepared in phosphate buffer at pH 7.5 (being used as coupling reagent) for 2-3 minutes.
(6) Wash in distilled water and mount in glycerine jelly.

Results: The development of blue or purple coloured sites indicate the presence of α-galactosidase activity.

(F) CHOLINESTERASE

Cholinesterase is the enzyme that hydrolyses acetylcholine to choline and acetic acid. Cholinesterases are differentiated into acetylcholinesterase (true) and cholinesterase (pseudo).

Experiment No. 1

Histochemical detection of cholinesterase in animal tissue using Acetylcholine method (Koelle and Friedenwall, 1949).

Solutions required

Solution A Dissolve 3.75g of glycerine in 18ml of 1N potassium hydroxide. Make the volume upto 100ml with distilled water.

Solution B 0.1M solution of cupric sulphate in distilled water.

Solution C Dissolve 14.5mg of acetyl thiocholine iodide in 0.75ml of distilled water. Add 0.25ml of solution B. Cupric iodide is precipitated leaving the acetyl thiocholine in the solution. Centrifuge and decant, retaining the clear solution.

Solution D Dissolve acetyl thiocholine iodide in a 1% solution of copper glycinate (to give an excess of copper glycinate). Leave this solution to overnight. Collect the precipitate (that of copper thiocholine), wash it several times with water.

E. INCUBATION MEDIUM

Distilled water	8.6ml
Solution B	0.2ml
Solution A	0.4ml

To this, add a trace of copper thiocholine. Heat to 37°C. Just before use, add 0.8ml of solution C. Filter on to the sections, *i.e.* it must be used immediately.

Procedure

(1) Prepare fresh frozen cryocut sections (8μ-10μ) and transfer them in a cavity block.

(2) Immediately pour the freshly prepared incubation medium and leave it for incubation at 37°C for 15-60 minutes depending on the nature of the tissue and thickness of the sections.

(3) Thoroughly wash with distilled water containing traces of copper thiocholine (only just sufficient quantity).

(4) Immerse the sections into yellow ammonium sulphide solution (just as in case of alkaline phosphatase test).

(5) Wash in distilled water.

(6) Mount in glycerine jelly.

Control: Any inhibitor, such as eserine, SO-OMPA or BW-284C5/ or 62C-47 at very low concentration should be added to normal incubation medium to inhibit the enzyme activity.

Results: Brown deposits often of crystalline nature indicate the presence of cholinesterase or acetylcholinesterase activity.

Experiment No. 2

Histochemical detection of cholinesterase in tissues using modified lead thiocholine method (Joo *et al.*, 1949).

Solution required

Incubation Medium

Acetylcholine	5.0mg
1.36% sodium acetate solution	2.5ml
0.6% acetic acid	0.3ml
7.0% lead nitrate solution	0.5ml

Mix in the given order.

Procedure

(1) Prepare frozen sections (8μ-10μ) of the tissue fixed in 10% formalin (preferably neutral) and transfer them in a cavity block.

(2) Directly pour the incubation medium on the free-floating sections and incubate them at 40°C for 10-15 minutes. (The incubation medium should be already at the required temperature before preparing the sections).

(3) Wash well in distilled water.

(4) Treat with freshly prepared 1% yellow ammonium sulphide solution for 1-2 minutes.

(5) Wash quickly in distilled water.

(6) Dehydrate in ascending grades of alcohol, clear in xylol and mount in DPX or in thin Canada balsam.

Results: The sites showing black deposits will correspond to the presence of cholinesterase activity.

Experiment No. 3

Histochemical detection of cholinesterase in animal tissue using modified Thiocholine method (Gerebtzoff, 1959).

Solutions required

A. 0.1M acetate buffer pH 5.0-6.2

B. acetyl choline iodide 15.0mg
cupric sulphate 7.0mg
distilled water 1.4ml
centrifuge and use supernatant fluid

C. 3.75% aqueous solution of glycine.

D. 2.5% aqueous solution of cupric sulphate

E. *Incubation medium*

Mix the above in the order given below:

Solution A	5.0ml
Solution B	0.8ml
Solution C	0.2ml
Solution D	0.2ml
Distilled water	3.8ml

Procedure

(1) Prepare cryocut frozen sections (8μ-10μ) of formol calcium fixed tissue and transfer them in a small cavity block.

(2) Rinse the sections in water to remove the fixative.

(3) Incubate the sections in incubation medium at 37°C for 20-80 minutes.

(4) Wash with two changes of distilled water, each time for 2-3 minutes.

(5) Treat with 1% freshly prepared solution of yellow ammonium sulphide for 1-2 minutes.

(6) Wash well with distilled water.

(7) Wash with running tap water and mount in glycerine jelly.

Results: The sites staining brown indicate areas exhibiting cholinesterase activity.

5

Paper Chromatography

INTRODUCTION, PRINCIPLES AND TYPES

The term chromatography was coined by the Polish Botanist Tswett in 1906 to represent a method used in separating the mixture of coloured compounds, such as pigments from each other. In pilot experiment, he observed that if a mixture of pigments in a solvent mixture of petroleum ether was percolated from the top of a narrow glass column that was firmly packed with $CaCO_3$, the pigments were separated from each other and appeared as coloured zones along the length of the column at different distances. He ascribed the differences in the behaviour of pigments in differing adsorbability, the more strongly adsorbed chlorophylls for example displacing weakly adsorbed xanthophylls and thus appearing further along the length of column under the influence of flowing solvent.

Nowadays, chromatography is defined as *a technique, which is used for the separation of closely related substances*, such as a mixture of amino acids or sugars. When the separation is carried out on paper, it is called ***paper chromatography***. The principle involved is that chromatography through a filter paper matrix occurs primarily as a result of partition of solute between two solvent phases, one mobile organic phase and the other bound to cellulose fibres and through physical adsorption and ion exchange. The use of paper (Whatman filter paper) for this purpose was described by Consdon, Gordon and Martin in the year 1944. This technique is based on the differential movement of different compounds of more or less similar nature at different speeds. Here, the paper acts as the inner support, as an adsorbent, as an ion exchange material due to its content of carboxyl groups. Cellulose fibres of paper have strong affinity for water present in the solvent. The aqueous phase (water), which is stationary is held up on paper. As the solvent flows across the paper having a solute, a partition of this solute occurs between the mobile phase (organic solvent/ organic phase) and the stationary phase (aqueous phase/ water) as a result of which a part of solute leaves the paper and enters the mobile phase. When the solvent reaches a zone of paper having no solute partition separation again takes place. Therefore, with the continuous flow of solvent, which causes partition between the two phases, the solute is transferred from point of its application to a point some distance away on the filter paper in the direction of solvent flow.

The simplest expression of the response of solute or substance to chromatography is denoted as ***'Rf value'***.

Rf value of solute is the physical constant and is defined as

$$Rf = \frac{\text{Distance solute travels from its point of origin}}{\text{Distance solvent front travels from its origin}}$$

The compounds are separated from each other on account of : (i) their flow at different rates on the porous surface, and (ii) their differential affinity to adhere on the porous surface.

1.1 Types of paper chromatography

The paper chromatography is meant for qualitative separation of different components in a given unknown mixture of amino acids, sugars, *etc.* It is of two types, namely,

(i) *Ascending paper chromatography*, and

(ii) *Descending paper chromatography.*

The principle of separation involved in both remains the same except for the difference that in case of ascending type, the direction of flow of solvent is vertically upwards, *i.e.* from bottom to top. The other apparent difference between the two is in the timings. Besides ascending and descending types, there is one more, *two dimensional paper chromatography.*

1.2 Paper Chromatography Apparatus

The paper chromatography is carried out in an apparatus called, chromatography chamber. The chromatography chambers most commonly used in the laboratories are either large airtight wooden chambers with glass windows and other supporting accessories (Figure 5.1) or cylindrical glass chamber (Figure 5.2).

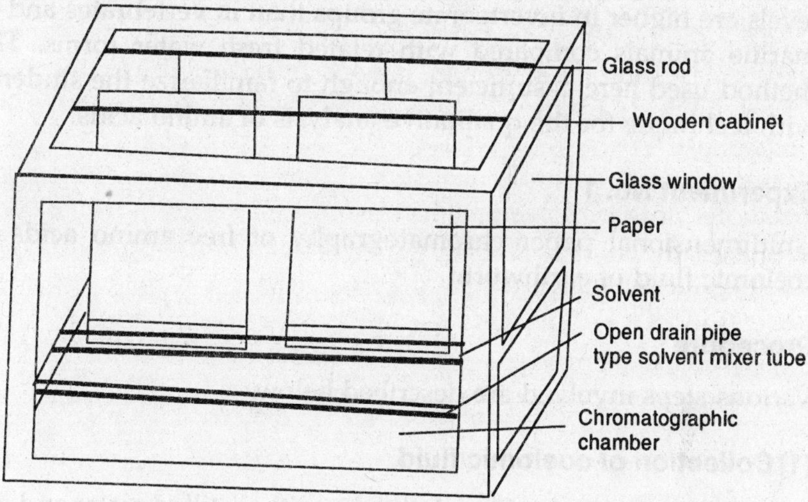

Fig. 5.1. Wooden chromatography chamber (ascending type)

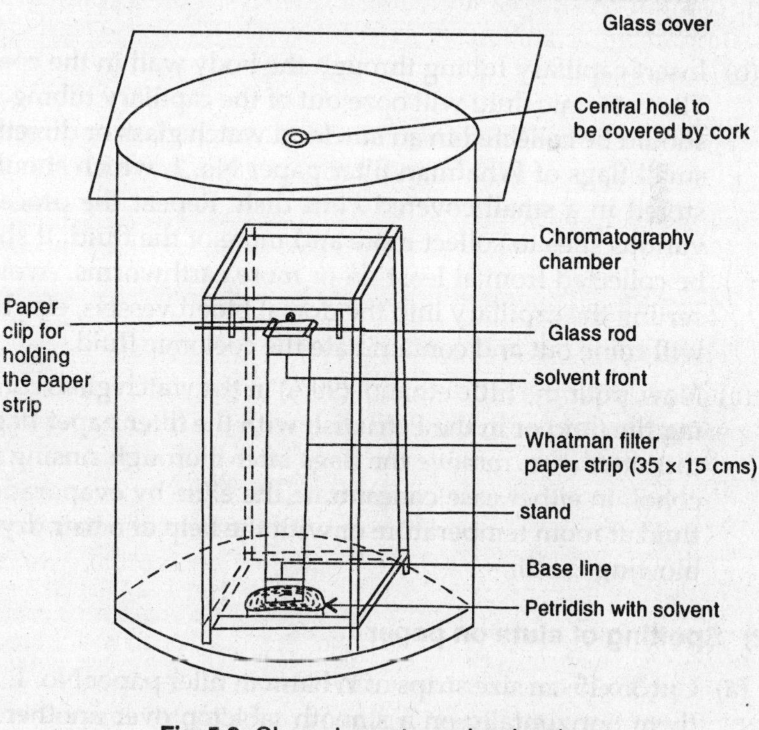

Fig. 5.2. Glass chromatography chamber

Occurrence of amino acids in animal tissue

A wide range of animals can be used according to the local availability. The free amino acid content of animal tissue varies greatly between the phyla and the classes within them. Generally, the levels are higher in invertebrate groups than in vertebrates and in marine animals compared with related fresh water forms. The method used here is sufficient enough to familiarize the students with technique for the qualitative analysis of amino acids.

Experiment No. 1

Unidimensional paper chromatography of free amino acids in coelomic fluid of earthworm.

Procedure

Various steps involved are described below.

(1) Collection of coelomic fluid

(a) Take a live animal, wash it twice with distilled water and dry it with blotting paper. Handle the worm gently during this operation as they discharge the coelomic fluid on slight irritation.

(b) Insert capillary tubing through the body wall in the coelom. The coelomic fluid will ooze out of the capillary tubing. This should be collected in an air-dried watch glass or directly on small flags of Whatman filter paper No. 1, which should be stored in a small covered Petri dish. Repeat the process at various sites to collect more and more of the fluid. It should be collected from at least six or more earthworms. Avoid inserting the capillary into the dorsal blood vessels, else blood will come out and contaminate the coelomic fluid.

(c) Now, pour the little ethanol (90%) in the watch glass containing the fluid or in the Petri dish with the filter paper flags. In the latter case, remove the flags after thorough rinsing in alcohol. In either case concentrate the elute by evaporation of fluid at room temperature or with the help of a hair dryer by blowing the air.

(2) Spotting of elute on paper

(a) Cut 35x15 cm size strips of Whatman filter paper No. 1. Place them horizontally on a smooth tabletop over another filter paper sheet. Draw a line with a pencil applying least pressure, along the width of the filter paper 3 cm away from one end.

(b) Make a circle with pencil on the line at the place where the spot of the test solution has to be applied. Now, take a capillary tubing (1mm internal diameter), dip it in the elute and transfer it on the circle. Dry it with hair dryer. Repeat the process till maximum quantity of elute has been transferred. The spot should be applied at a sufficient height from the bottom edge, so as to avoid the immersion of the compounds in the solvent.

(c) In the same way as above, place the spots of the known samples separately on pencil line and label each spot. Store these strips in the neat, clean and dry place. When the spots are completely dry, the strip is ready for further processing.

(3) Preparation of the solvent and the tank

Prepare the mixture of n-butanol, acetic acid and distilled water in the ratio of 4:1:1 (v:v:v) in a separating funnel and shake well. This is left to stand for a few hours. The lower layer is then run off and an aliquot of lower layer is placed in a Petri dish at the bottom of an airtight tank or chromatographic chamber to saturate its internal atmosphere with solvent vapours.

(4) Development

The end of the paper chromatography strip bearing dried spots is placed in a trough kept inside the airtight chromatography chamber. This should be carried out as quickly as possible to avoid unsaturation of chamber. It is secured in the place with a rod or by means of a thread. Enough of solvent mixture is then poured in the trough and the chamber is then closed. The paper strip is left as such for an overnight period or for that much of the period till the solvent front reaches to a predetermined mark. After the run, the strip is taken out and the solvent front is marked with a pencil. This should be roughly 30-32cm from the base line. Hang the strip to dry in the air. The drying may be hastened by using a hair dryer.

Further development is carried out by spraying/dipping the strip with 0.1 to 0.25% solution of Ninhydrin (indane-trione hydrate) in butanol or butanol water. Sometimes 2% Ninhydrin solution is also used. The strip is again allowed to dry at room temperature aided with hair dryer. This results in the development of coloured spots of different amino acids at different distances within 30-40 minutes. Make the outline of each spot with a pencil. Identify these amino acids by comparing them with the spots of known amino acids.

The spraying with colour developing reagent should be fine and uniform.

Calculation of Rf value

Measure the distance travelled by the solvent, *i.e.* the distance between the line on which the spots of amino acid sample were applied and the solvent front. Mark the spots as 1, 2, 3, and so on by encoding them and measure the distance of flow in centimetres for different amino acid spots. Each time measure the distance from the base line to the centre of the developed spot and calculate Rf value for each spot as below:

$$Rf = \frac{\text{Distance of flow of individual amino acid}}{\text{Distance of solvent front}}$$

By comparing the Rf values given in the book or from the spots (Rf value) of known amino acids, one can find out the total number of amino acids present in the mixture.

Result: The unknown sample contains the following amino acids: 1. ------------------- 2. ------------------- 3. --------------

An illustration of the chromatogram developed this way is given in Figure 5.3.

Precautions

(1) The chromatography chamber should be kept airtight in order to make it saturated with vapours of the solvent.

(2) Only distinct spots should be marked with pencil.

(3) Spot of each sample should be applied in very small quantity.

(4) For spotting, capillary tubings of uniform diameter should be used.

(5) Base lines carrying spots should under no circumstances dip in solvent mixture.

(6) The strip should remain in almost vertical conditions when solvent is allowed to ascend or run upwards.

(7) Contamination of filter paper by sweat or perspiration must be avoided as they contain amino acids.

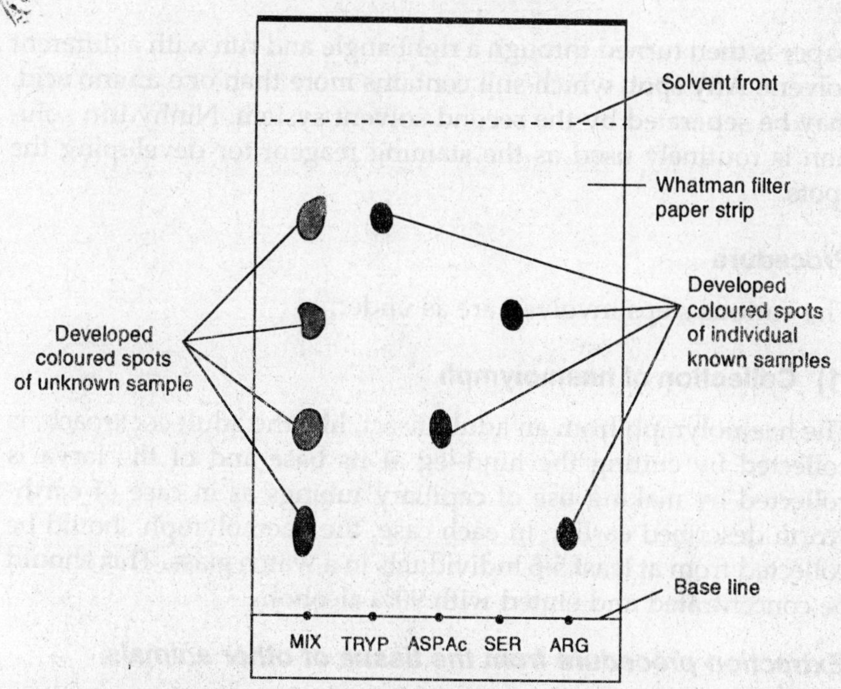

Fig. 5.3. Developed chromatogram of amino acids

MIX = Mixture of unknown amino acids; TRYP = Tryptophan; ASP.Ac = Aspartic acid; SER = Serine; ARG = Arginine

(8) The concentration of amino acids separated or partitioned should be in the range of 5-15 μg of each amino acid in 2.4μl of solution.

Experiment No. 2

Two Dimensional paper chromatography for the quantitative analysis of free amino acids in the haemolymph of mature adult insect or larvae.

Theory: Two-dimensional paper chromatography is superior to unidimensional paper chromatography, as there are a variety of solvents used in paper chromatography, which are never equally suitable for all amino acids. It is, therefore, necessary to use several different solvent systems and a suitable way of doing so is to carry out two dimensional paper chromatography.

In this technique, a squarish paper sheet of Whatman No. 1 filter paper, measuring preferably 34 x 34 cm in size is preferred instead of a paper strip. The spots applied are first allowed to run in one direction with a given solvent mixture and after the first run, the paper is then turned through a right angle and run with a different solvent. Any spot, which still contains more than one amino acid, may be separated by the second solvent system. Ninhydrin solution is routinely used as the staining reagent for developing the spots.

Procedure

The various steps involved are as under:

(1) Collection of haemolymph

The haemolymph from an adult insect, like the adult cockroach, is collected by cutting the hind-leg at its base and of the larva is collected by making use of capillary tubings as in case of earthworm described earlier. In each case, the haemolymph should be collected from at least 5-6 individuals in a watch glass. This should be concentrated and eluted with 90% alcohol.

Extraction procedure from the tissue of other animals

The procedure followed here is meant for amino acids pool of the haemolymph of cockroach. This is then extracted with ethanol. Thereafter the elute extract is concentrated. Besides the extracts of many other animal tissues can be prepared and used, e.g. remove separately 1-2g of the rat liver and/or leg muscle, crab or crayfish hepatopancreas and/or leg muscles (1 or 2 animals required), locust or grasshopper thoracic muscle (3 or 4 animals required). Blot gently on filter paper and place in 2 volumes of 80% ethanol in a test tube, bring to boil, then homogenate for 2-3 minutes in a homogenizer. Leave the extract for ½ hour, then centrifuge for 30 minutes. Use the supernatants for chromatographic analysis or else store in cold room in glass stoppered vessels till used.

(2) Spotting of sample

Cut out a chromatographic paper 34 x 34cms dimension from Whatman filter paper No. 1. Draw a line at a distance of about 3cms from its lower edge and another line at the same distance from the edge of a right hand side of paper at right angle to the first line. Mark a point where the two lines meet each other. Mark the direction of the flow of the two solvent systems (butanol: acetic acid: water, 4:1:1, v:v:v and phenol: water, 1:1, v:v:v) along two lines. Put a spot of the elute with a capillary, at a point where the two lines meet each other. Repeat the process several times till enough of elute is transferred on the spot.

(3) Preparation of the chamber

Place the solvent in trough and put it inside the chamber so that the atmosphere inside it is saturated. The solvent system consisting of butanol: acetic acid: water in the ratio of 4:1:1, v:v:v should be used here.

(4) Development

Roll the paper into a cylinder by sewing loosely with a white thread. Place the chromatogram in the solvent contained in the trough and ensure that the chamber is air tight. Allow the solvent to run till it has travelled a considerable distance or has reached to a predetermined mark. Take out the chromatogram and mark the solvent front with a pencil. Allow it to dry in the air or with the help of a hair dryer at room temperature.

Remove the thread from the paper and again roll it into a cylinder but in the direction at right angles to the first and place it as before in the second solvent system consisting of phenol: water. Allow the solvent to ascend to a suitable distance. Take out the chromatogram and mark the level upto, which the solvent had travelled. Allow the chromatogram to dry.

(5) Developing of spots

Locate the amino acids by spraying the dried chromatogram with 2% ninhydrin solution in butanol.

(6) Identification of amino acids

The identification of separated amino acids is carried out either by calculating their Rf values or by comparing them with those of the spots of known amino acids on standard chromatograms run under identical conditions.

For preparing the standard chromatogram for identification purposes, a marker mixture of the amino acids is run under the same chromatographic conditions as used for the body fluids or tissue extracts. An illustration of the separation of a marker mixture achieved is shown in Figure 5.4.

Common solvent system for two dimensional chromatography

(1) (a) Butanol-acetic acid-water, 4:1:1, (v:v:v)

(b) m. cresol-phenol (pH. 9.3 borate buffer-water (1:1, v:v:v).

Experiment No. 3

Unidimensional paper chromatographic analysis of sugars.

Theory: In case of separation of sugars by paper chromatography the filter paper also on account of the presence of carboxyl groups in its composition acts as an ion-exchange material.

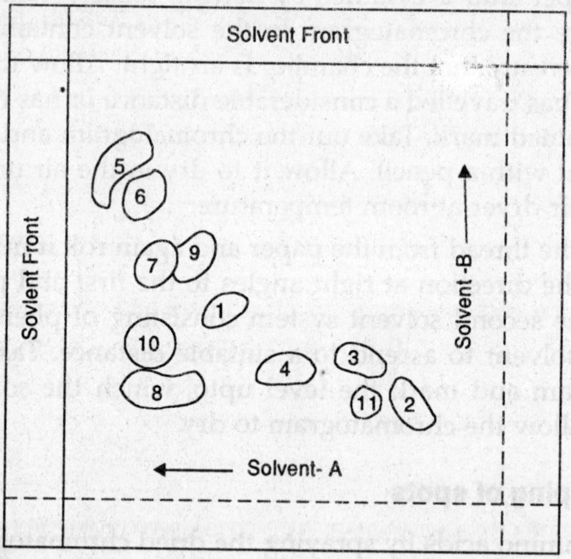

Fig. 5.4. Pattern of spots from two dimensional paper chromatography of a marker solution containing (1) alanine, (2) aspartic acid, (3) glutanic acid, (4) glycine, (5) leucine, (6) valine, (7) proline, (2) lysine, (9) tyrosine, (10) arginine and (11) taurine. solvent A. n-Butanol: acetic acid: water; Solvent B. Phenol: water.

On account of a strong affinity of cellulose fibres, constituting the paper, for water in the solvent system, the aqueous phase (water), which acts as the solvent across the paper, a partition of the dissolved solute occurs between the mobile/ labile phase (organic solvent/ phase) and the stationary phase (aqueous phase/water). As a result, a part of the solute leaves the paper and enters the mobile phase. When the solvent reaches a zone of paper having no solute, partition again occurs. With the continuous flow of the solvent, which is responsible for the partition between two phases, the solute is transported from point of its origin to a point some distance away on the filter paper in the direction of the solvent flow.

In this case the simplest expression of the response of a solute or a substance in chromatography is expressed as Rf values.

Procedure

The various steps involved are as under:

(1) Preparation of Benzidine reagent (Spot developing reagent)

This is prepared by dissolving 500g of Trichlor acetic acid (TCA) in 5ml of distilled water and 90ml of acetone. To this mixture is added just sufficient quantity of water to make the volume of solution to 1000ml.

(2) Spotting of test and standard samples

Cut Whatman No. 1 filter paper sheets into strips of 4" x 20". Using a lead pencil, draw a transverse base line 1" away from the lower end of the paper strip. Mark six or seven spots at equal distance with lead pencil and specify the names of known standard sugar samples and unknown mixture under analysis below the specific spots.

Now, lay the strips on a flat tabletop and with the help of capillary tubings of approximately uniform dimensions, apply the specific sugar solutions against the spot marked for it. This is carried out by first dipping the capillary in the sample to be applied on the spot. Gently touch the tip of the capillary tubing on the spot. Repeat the process several times till just sufficient quantity of solution is not transferred or loaded on the spot. While loading the spot go on drying it with the help of hair dryer. The capillary tubing at the time of application on the spot should be held in upright position. Take care that the diameter of the spot bearing the samples does not exceed for more than 0.3μ. This way apply the solutions of unknown sample at one spot. Every time using separate capillary tubing for each sample.

(3) Development of chromatogram

These strips bearing the dried spots are developed in the ascending manner in a chromatographic chamber in the solvent system consisting of butanol-acetic acid-water in the ratio of 4:1:1.

For this, the atmosphere inside the chromatographic chamber is first saturated with the vapours of the solvent system. This is achieved by placing some solvent in a large Petri dish inside the chamber, which is then left for sometime with thread in such a position so that their lower ends dip in the solvent. Care should be taken so that the spots do not dip in the solvent. Leave it as such for an overnight period. When the solvent has travelled to the desired mark or height, take out the chromatograms from inside the chamber and allow them to dry up at ambient temperature. This drying of the chromatograms can be hastened by use of hair dryers, through which the air is blown at room temperature. Mark the solvent front. An illustration of developed chromatogram is given in Figure 5.5.

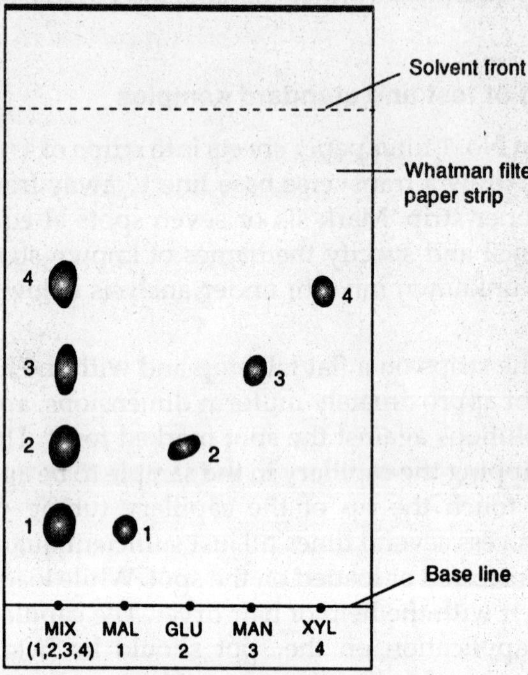

Fig. 5.5. Replica of ascending paper chromatograph of sugars.
MIX = Mixture of unknown sample
MAL = Maltose; GLU = Glucose
MAN = Mannose; XYL = Xylose

(4) Developing the spots

Dip at least twice or thrice the dried chromatogram in Benzidine reagent for 2-3 minutes and allow it to dry in the air at room temperature for a few minutes and thereafter, in hot air oven or electrical incubator set at 110^0C for 3-5 minutes. All sugar spots will be rendered yellowish brown against a light yellow or dull black background.

(5) Evaluation of results

By comparing the spots of sugars in the unknown sample with that of the spots of the sugars of known standards, find out the sugars present. Also find out Rf values of the spots developed and by comparison with Rf values of the known standards given in literature developed under identical conditions, find out the sugars present.

Precautions

(1) Chromatographic chamber should always be airtight.

(2) Avoid as far as possible, the contamination of the chromatographic paper with perspiration or sweat of hand.

(3) Keep the capillary tubing in an upright position at the time of applying the spots of either the unknown or known samples. The diameter of the loaded spot should not exceed for more than 0.3μ.

(4) Dipping of the developed chromatogram in the Benzidine reagent should be done quickly and uniformly. Care should be exercised in avoiding the direct contact of Benzidine solution with the skin, as it is highly carcinogenic, *i.e.* cancer producing.

(5) For each sample separate tubing should be used.

6

Thin Layer Chromatography

INTRODUCTION AND PRINCIPLE

Thin layer chromatography (TLC) is a simple, rapid and efficient method of separating different substances including amino acids.

The principle involved in the separation of amino acids by this technique is more or less the same as in case of paper chromatography, except that in place of Whatman filter paper a thin uniform layer of silica gel G borne on a glass plate is used, which acts as an inert support, an adsorbent and an ion exchange matrix.

The simplest expression of the response of a solute or substance to TLC is also denoted as Rf value, as already mentioned in the previous chapter. Rf value of a solute is the physical constant of that solute at particular temperature for a specific solvent mixture.

A. THIN LAYER CHROMATOGRAPHY OF AMINO ACIDS

Apparatus, reagents, *etc.*

Glass plates (7 x 4inch in size), cylindrical glass troughs (9 x 6inch) with a glass lid to cover the top, Petri dish, silica gel with plaster of Paris as binder (silica gel G, E. Merck), fine capillary tubings, temperature bath at 110°C, chromogenic reagent for staining amino acid spots. The cadmium acetate solution is stored as stock solution and ninhydrin is added at the required time for use.

Preparation of Chromogenic reagent (Staining solution)

It is prepared by adding 10-12ml of cadmium acetate solution in 0.25g of ninhydrin dissolved in 100ml of acetone. The cadmium acetate solution itself is prepared by mixing 900ml of distilled water, 180ml of glacial acetic acid and 9gm cadmium acetate. This should be used as stock solution and ninhydrin should be added as and when required.

Solvents for separating amino acids

Any of the following solvent system can be used for separating the amino acids.

(1) Ethanol (96%)/water (63: 37; v/v)

(2) n-Propanol / water (64: 36; v/v)

(3) n-Butanol/ acetic acid/ water (60: 20: 20; v/v/v)

Rf values of the most common amino acids are displayed in the Table 6.1.

Table 6.1: Rf values of amino acids

S. No.	Name of amino acids	Rf values in solvent
1.	Alanine	0.45
2.	Arginine	0.06
3.	Aspartic acid	0.56
4.	Cysteic acid	0.65
5.	Cystine	0.39
6.	Glutamic acid	0.56
7.	Glycine	0.42
8.	Histidine	0.30
9.	Leucine	0.60
10.	Lysine	0.04
11.	Methionine	0.58
12.	Norvaline	0.48
13.	Phenyl alanine	0.62
14.	Serine	0.48
15.	Theonine	0.51
16.	Tyrosine	0.67
17.	Valine	0.58

Procedure

(1) Preparation of the chromatoplates

Thoroughly wash the glass plates with detergent and immerse for several hours in sulphuric and chromic acid mixture. Thereafter, wash the plates with tap water and finally with distilled water. Leave the cleaned plates on a rack at room temperature to dry up. Arrange these plates on a large glass plate. These plates should be placed horizontally in a row, with edges in alignment. Fasten both the ends of all plates with cello tape as shown in (Figure 6.1). This will not only provide anchorage to the plates but will also ensure a film of uniform thickness.

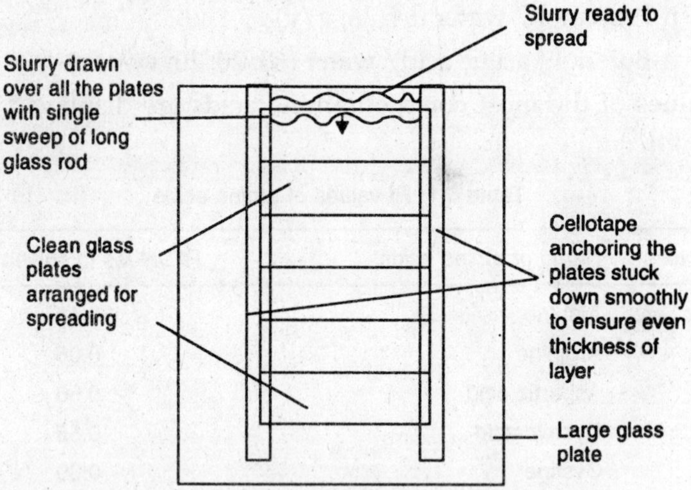

Fig. 6.1. To illustrate (1) arrangement of plates, and (2) spreading of slurry

Prepare the slurry of silica gel by mixing 25grams of silica gel G powder with 50ml of distilled water in a 150ml conical flask fitted with a clean stopper followed by vigorous shaking for at least 30seconds. For preparing TLC plates having a thickness of 0.5mm silica gel G, transfer the thin paste (slurry) on plates and immediately spread it with a spreader by puddling it across the row of plates with uniform motion not stopping until all the plates were coated. Any time, if wrinkles appeared in the layer, each plate should be taken while still wet and its edge should be gently bumped on a smooth horizontal surface for uniform layer. The spreader used usually consists of a glass rod.

Dry the prepared chromatoplates at room temperature and activate them at 110°C in a hot air electric oven for 1 ½ hour. Store these plates in a cabinet, which should contain "Blue" silica gel as a dessicant.

(2) Preparation of the Tank

If the glass tank specially designed for TLC is not available then use either a cylindrical glass trough of 6" diameter in width and 9" in height with lid or beaker of equal dimension. The so called tank should be internally lined with Whatman No. 1 filter paper and pour the solvent mixture into a Petri dish already kept at the bottom of the tank. Besides pouring the solvent mixture in Petri dish, also pour some quantity at the bottom of tank so that the lower end of the filter paper is wetted. The volume of the solvent mixture at the bottom should not be less than ¼ of an inch. Filter paper lining will ensure the complete saturation of the tank. The tank is covered with either the glass plate or with aluminium foil seemed in place with a rubber band. The complete saturation of the tank takes place in 30 min to 2 hours depending on the solvent mixture used in Figure 6.2.

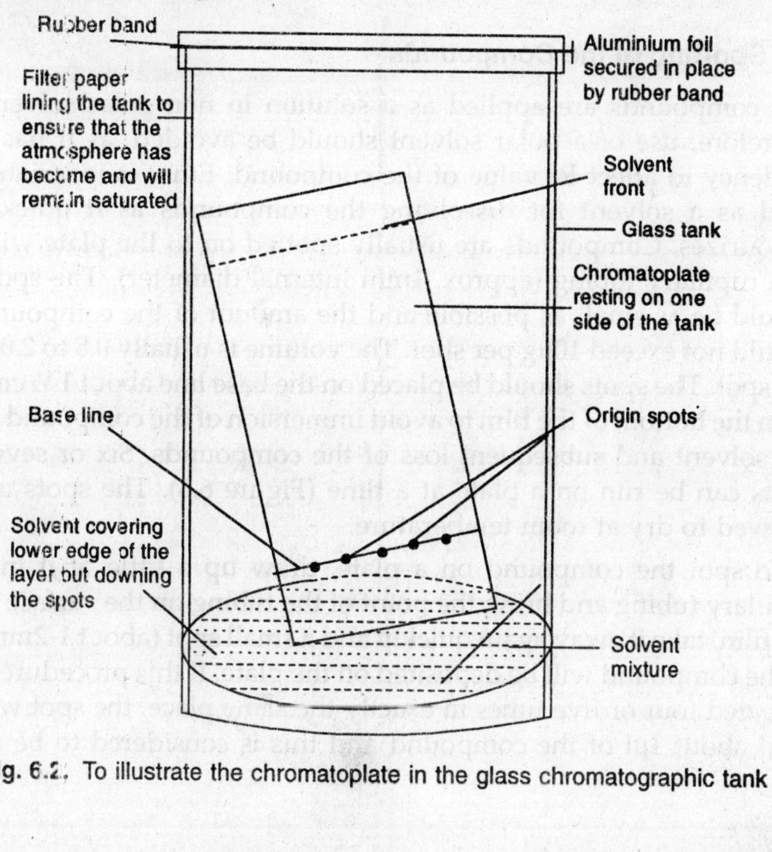

Fig. 6.2. To illustrate the chromatoplate in the glass chromatographic tank

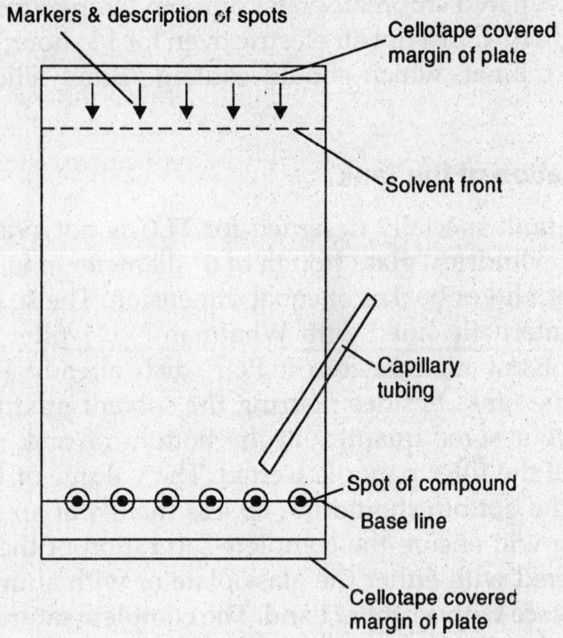

Fig. 6.3. To illustrate appearance of a plate during and after spreading

(3) Spotting of the Compounds

The compounds are applied as a solution in non-polar solvent. Therefore, use of a polar solvent should be avoided, as it has a tendency to affect Rf value of the compound. Ether is invariably used as a solvent for dissolving the compounds as it quickly vapourizes. Compounds are usually spotted on to the plate with thin capillary tubing (approx. 1mm internal diameter). The spots should be as small as possible and the amount of the compound should not exceed 10μg per spot. The volume is usually 0.5 to 2.0μl per spot. The spots should be placed on the base line about 1 ½ cms from the bottom of the film to avoid immersion of the compound in the solvent and subsequent loss of the compounds. Six or seven spots can be run on a plate at a time (Figure 6.3). The spots are allowed to dry at room temperature.

To spot the compound on a plate, draw up a little of it in a capillary tubing and bring the point of the tubing on the surface of the film, take it away again quickly and a small spot (about 1-2mm) of the compound will be deposited on the plate. If this procedure is repeated four or five times in exactly the same place, the spot will hold about 1μl of the compound and this is considered to be an ideal amount for accurate resolution. Before trying the next application, make sure that each of the application to the spot is quite dry. If this practice is not strictly followed, the spot will spread and the resolution will either be lowered or will not be significantly accurate.

(4) Development of plates

When each spot is completely dry, the plate is ready for development. The plate is placed in a glass chamber containing a suitable solvent or a solvent mixture (depending upon the nature of the compound to be studied). The chamber is closed by placing the glass lid and the solvent is allowed to run till it did not reach the finish line or a predetermined mark. Every precaution should be taken to avoid the travelling of the spot beyond the finish line and whenever it is found to be so, the solvent mixture should be accordingly modified. After the run, the plates developed in solvents having higher boiling points should be dried at 110°C in a hot air oven, while those developed in solvents having low boiling points should be dried at room temperature. The plate should be placed in the developing tank quickly else there is every likelihood that unsaturation of tank atmosphere may take place.

(5) Visualization of spots

For visualizing the spots, the dried plate is sprayed with colour developing reagent consisting of ninhydrin in cadmium acetate solution. For colour development, place plate at 110°C for 10 minutes (Figure 6.4).

(6) Calculation of Rf values

Rf values should be calculated by preparing the replica of the fully developed chromatoplates. Comparing the calculated Rf values with those given in the table 6.1 should carry out the identification.

Precautions

There are actually three factors, which usually affect the Rf value of a separated compound.

(1) A chromatoplate, which does not carry an evenly spread film, is likely to yield much lower Rf values in comparison to a film of even thickness. Likewise, an even film which is thicker than 20μ will reveal the Rf values too low.

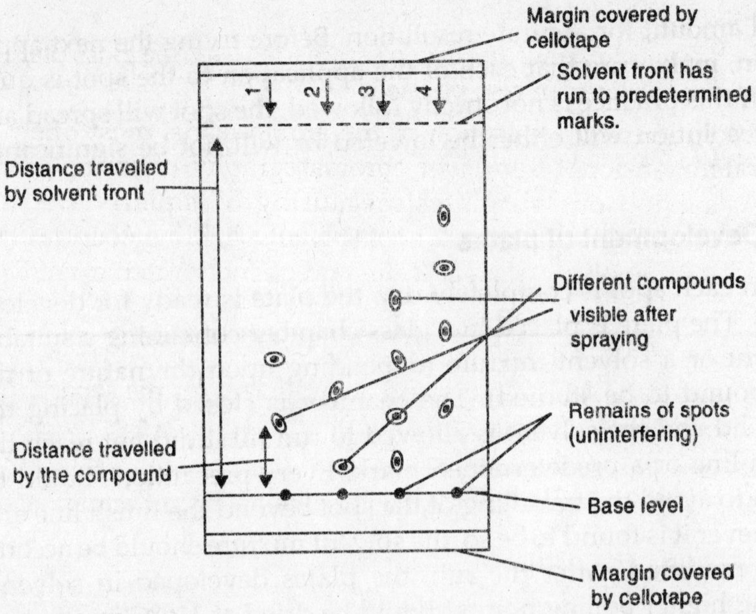

Fig. 6.4. To illustrate a model chromatoplate after developing and spraying

(2) The saturation of the chromatography chamber also affects the Rf value of a compound. When the atmosphere of tank is not fully saturated, the solvent will go up the film and evaporate. Therefore, more solvent will move through the film to carry the solvent front to the pre-determined mark than if the atmosphere was saturated. Hence, the Rf value will be significantly higher than that obtained under optimal conditions. The greater source of error in the measurement of Rf value is the over-development of a chromatogram.

(3) If the maximum load per spot exceeds 10µg, the spot will become larger and Rf value will increase progressively.

(4) The extract of animal tissue before being analyzed for amino acids should be desalted. Treating the extract with 95% alcohol, which takes up the amino acids above and leaves the salts behind, best carries out the desalting.

(5) The spraying of the colour developing reagent/solution should be done with either a glass automizer or specially designed spray bottles because the use of both these ensures a fine, uniform spray. Harsh spraying should always be avoided, as it will blow the powder off the plate. Care should be taken to ensure that the reagent is evenly distributed throughout the entire surface area of the plate.

Note:

(1) If the number of spots to be applied is less, the glass plate of 7 x 2 ½″ size can also be used.

(2) For perfect separation of the amino acids, most often two dimensional thin layer chromatography of amino acids is carried out on plates measuring 6″ square. The most appropriate combination of solvents yielding good results is *n*-butanol: acetic acid: water and phenol: water: cyanide.

B. THIN LAYER CHROMATOGRAPHY OF SUGARS

This technique is a simple, rapid and efficient for the separation of different sugars. The principle involved is more or less the same as in case of TLC of amino acids.

Apparatus, reagent, *etc.* All as in case of TLC of amino acids.

Procedure

The chromatoplates are prepared in the usual way but they are pre-run in 0.02M sodium acetate solution. When the film is saturated with sodium acetate solution, the chromatoplates are again dried at room temperature.

Table 6.2: Rf values of different sugars separated in isopropanol: ethyl acetate : water (52: 25:14; v/v/v) solvent mixture.

S. No.	Sugar	Rf value
1.	Arabinose	0.202
2.	Fructose	0.213
3.	Galactose	0.164
4.	Glucose	0.175
5.	Lactose	0.046
6.	Maltose	0.057
7.	Mannose	0.198
8.	Rhamnose	0.599
9.	Ribose	0.4210
10.	Xylose	0.35

The spots of the samples are loaded as in case of amino acids. These are then dried. The chromatography chamber is saturated with vapours of the solvent mixture consisting of ethyl acetate/ isopropanol/water in the ratio of 52: 28: 14 (V/V/V). The dried plates are then run in the same solvent mixture as usual. After the run, the plates are taken out and dried. The spots of separated sugars and of known samples are detected by spraying the dried plates with anisaldehyde solution, which consists of 9ml ethanol (100%), 0.5ml concentrated H_2SO_4 and 0.5ml anisaldehyde. The development of coloured spots takes place when the sprayed plates are placed at 100°C for about 20 minutes.

7

Electrophoresis

INTRODUCTION AND PRINCIPLE

Electrophoresis is the name of phenomenon of the movement of small charged particles, usually ions through an electrolyte subjected to an electric field. If these are differently charged, they will move in opposite direction, the positively charged migrating to the cathode and the negatively charged to the anode. The rate of migrations of the particles/ions of like charges will depend among other things on the number of charges each carries. As a result of different rates of migration a complex mixture, such as plasma proteins, can be separated into a number of fractions of like mobility, the sharpness of separation depending upon the extent to which each fraction is homogenous in mobility.

Electrophoresis has been used for the separation of small ions, for example amino acids, but the process is similar to that for the separation of proteins and it is customary to use the term ***electrophoresis*** in all cases. The first technique to be widely used was that of Tisellius (1937).

By means of a very simple experiment detailed below (Figure 7.1) the signs of the charge on a colloidal particle can be determined very easily. Take a U shaped glass tube of uniform diameter and place in its bottom a colloidal solution, which will be distributed equally in the lower portions of both the arms of U tube. Thereafter, carefully add distilled water in each arm and immerse two separate wires one in each arm dipping in the water and on the other end both connected to a battery of high voltage. On switching the current you will observe that the colloidal particles migrate to the pole of battery having a charge opposite to that of the colloidal particles. This will indicate that the colloidal particles migrating to the negative pole are positively charged.

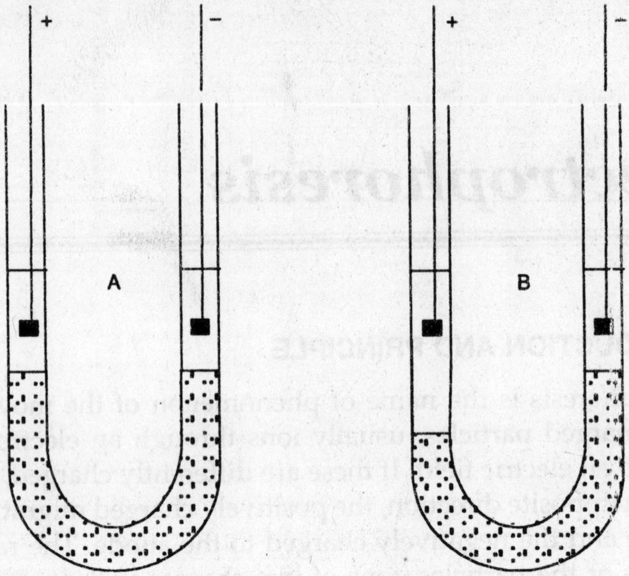

A. Before electrophoresis B. After electrophoresis

Fig. 7.1. Apparatus for demonstrating electrophoresis

The electrophoretic procedures are much simpler to perform, as much less material is required and many runs can be carried out at the same time. Filter paper was the first convection medium to be widely used and held the field for several years but numerous other media, such as, cellulose acetate, agar gel, starch block, starch-gel and polyacrylamide gel have been used subsequently and some applications have important advantages over others. But even after all these advantages, the paper electrophoresis and starch gel electrophoresis are still frequently used in the laboratories to apprise the students of this technique.

A. PAPER ELECTROPHORESIS

In this type, the supporting medium is a sheet of absorbent paper. The most satisfactory apparatus as used in laboratories are of two types according to whether paper is stretched horizontally or suspended over a support so that it forms an inverted V. In both cases, the process is carried out in a covered tank. These are thus named as: (1) Hanging strip inverted V type, and (2) Horizontal strip type.

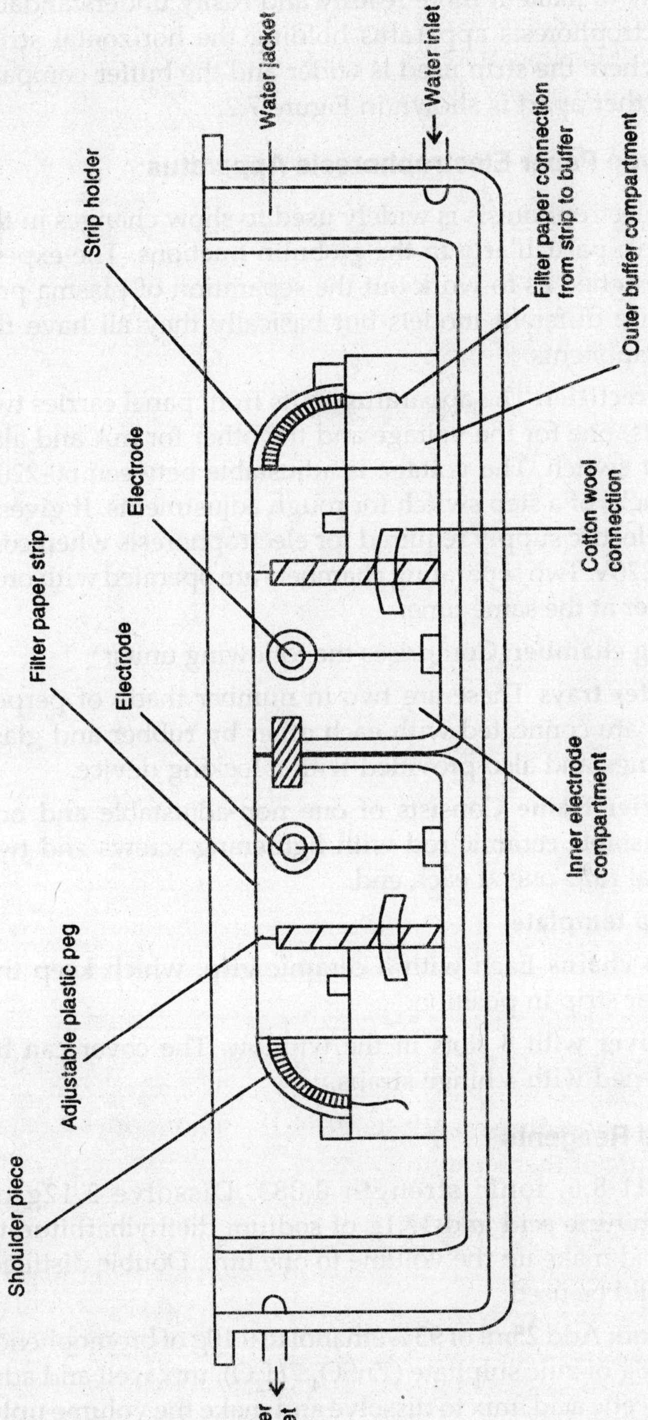

Fig. 7.2. Horizontal Strip Type Paper Electrophoresis Tank (After Shandon Universal Apparatus)

With a view to make it more readily and easily understandable a simple electrophoresis apparatus holding the horizontal strips type paper, where the strip used is wider and the buffer compartments are further apart is shown in Figure 7.2.

Horizontal Type Paper Electrophoresis Apparatus

This type of electrophoresis is widely used to show changes in the serum proteins, particularly in the globulin fractions. The experiment is so designed as to work out the separation of plasma proteins. There are different models but basically they all have the following components.

(1) **Constant rectifier:** The apparatus on its front panel carries two metre dials, one for the voltage and the other for mA and also one on-off switch. The voltage is adjustable between 60-220V with the help of a step switch for rough adjustments. It gives a constant electric supply required for electrophoresis when connected to 220V. Two separating chambers are operated with only one rectifier at the same time.

(2) **Separating chamber:** Comprises the following units:

   (a) **Buffer trays** These are two in number made of perpex and are connected with each other by rubber and glass tubings and also provided with a locking device.

   (b) **Carrier frame** Consists of one non-adjustable and one adjustable ceramic rod with tightening screws and two metal rails one at each end.

   (c) **Strip template**

   (d) **Two chains** Each with a ceramic grip, which keep the paper strip in position.

   (e) **A cover** with 8 slots in the window. The cover can be fastened with 4 hinge straps.

Solutions and Reagents

(1) **Buffer:** pH 8.6, ionic strength 0.083. Dissolve 3.12g of diethylbarbituric acid and 17.1g of sodium diethylbarbiturate in water and make up the volume to one litre. Double distilled water should be used.

(2) **Dye solution:** Add 25ml of 95% ethanol to 0.10g of bromophenol blue and 50g of zinc sulphate ($ZnSO_4.7H_2O$), mix well and add 5% (v/v) acetic acid, mix to dissolve and make the volume upto a litre with further addition of acetic acid 5%.

(3) **Wash solution:** Acetic acid, 5% (v/v).

(4) **Fixative solution:** Acetic acid, 5%(v/v) containing 0.3% of sodium acetate ($CH_3COONa.3H_2O$).

Setting up and operation of the apparatus

This is carried out step by step as under:

(1) Remove the cover from the separating chamber, lift out the carrier frame along with two ceramic chains.

(2) Pour sufficient quantities of buffer solutions in the buffer trays.

(3) Cut strips of Whatman 3MM filter paper of 40mm x 300mm and place the strip on the holder.

(4) Wet the strips thoroughly with the buffer and place the lid in position. For placing the strip in proper position, lift out the carrier frame along with two ceramic chains.

(5) Insert the carrier frame in the strip template so that the non-adjustable ceramic rod comes to lie in the recesses of strip template and adjust the rods to minimum distance using tightening screws. Place the strips properly in position on carrier frame with the help of ceramic chains.

(6) Pick up the carrier frame by metal rails at side and insert it in the separating chamber and immerse the projecting end of the strips in buffer trays. If strip drops then adjust with tensioning screws, release the locking device of the depression button. The paper is drawn out in horizontal position so that the apparatus is wider than the strip and buffer compartments are further apart.

(7) Put the cover with 8 slots in its proper position and fasten it with hinge strips.

(8) Leave the outfit covered in this position for 15-30 minutes for equilibrium to be reached, and the chamber becomes completely saturated. Since there is a smaller volume of atmosphere in the horizontal type this becomes saturated more quickly and the effect of heating is less marked.

(9) Depending on the width of the strips, apply 10μl of either one or more samples of serum through each slot separately and uniformly along a pencil line drawn about two-fifth of the length of paper from the end nearest to the negative pole with the horizontal type. In case of bigger paper strips or sheets, the amount to be applied on is adjusted accordingly.

(10) Connect the outfit or chamber with constant rectifier to 220 AC

mains, which supplies DC that is adjustable between 60-220 V by making use of step switch for rough adjustment and potentiometer for fine adjustments.

(11) Switch the current on and allow running for 16-18 hours. A voltage gradient of 3.2 volts and a current of 1.5 to 2.0mA is suitable for satisfactory operation.

(12) After electrophoretic run, take out the cover; remove the strip from the bath, by tearing off the projecting parts of the strip at bath ends. Remove the carrier frame and take out the strip and dry it at 100°C or 110°C for 30 minutes, preferably in a forced draught oven or a drying cabinet.

(13) Stain at room temperature for at least six hours by dipping the strip with strip holder in a staining tank. It is convenient to leave the strips staining overnight.

(14) After staining, wash for 6 minutes each time in two changes of the washing solution, then in the fixative solution for a further time period of six minutes.

(15) Blot the strips between clean filter papers and dry in a hot air oven at 100°C-110°C for about 15 minutes. Strips are then ready for examination. If necessary the blue colour can be developed further by exposing the strip to ammonia vapour. By comparison with known samples, a qualitative report can be prepared otherwise for quantitative assessment scanning and elution methods are used.

Observation and Results

The albumin moves most rapidly so it is found at the greatest distance from the start line. It is followed in turn by the α_1, α_2, β and γ-globulins. The γ-globulins have the smallest negative charge and since the movement due to this is approximately balanced by an opposite one due to electrophoresis they move but little, and with slight variations in the conditions of the experiment may be found at or slightly to either side of start line. The other fractions move towards the anode. Fibrinogen, when plasma is used, is found as a narrow band between β and γ-globulins. A scanned electrophoresis strip is shown in Figure 7.3.

Precautions

(1) Tougher papers like that of Whatman No. 3MM filter paper should be used as they are easy to handle also that trailing is more marked with such paper.

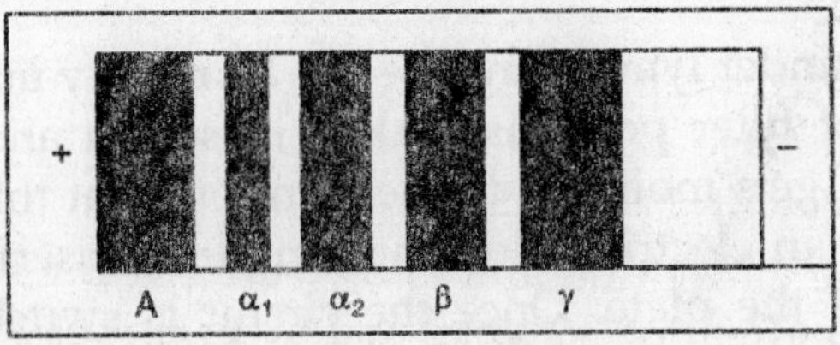

Fig. 7.3. Scanning of electrophoresis strip

(2) Buffers of ionic strength 0.08 to 0.10 appear to be more satisfactory. The rate of migration increases with decrease in ionic strength of buffer but bands become diffused.

(3) Great care is required for loading the sample; the most ideal one is 10µl of serum for a strip of one cm width.

(4) A time run of 16-18 hours is most convenient, safe and satisfactory.

(5) Since, there is some relationship between the drying and dye binding power of albumin, therefore, both these factors should be controlled very carefully.

(6) Irrespective of time-consuming factor, staining with weak bromophenol blue is preferable.

(7) The temperature of staining is not critical as the uptake of dye decreases with rise in temperature but it affects equally well to all.

(8) At pH 8.6 all proteins are anions and good separation is obtained. In case of proteins, the rate of migration increases as the pH of the buffer is moved away from the isoelectric point.

Practical Utility of Electrophoresis

The most significant use of this technique is in clinical studies where in minimum possible time a large number of samples can be subjected to analysis at a time, particularly in view of the fact that such fractionations show interesting variation in size of globulin fractions in certain diseases.

B. STARCH AND GEL ELECTROPHORESIS OF RAT SERUM

The details presented here are primarily based on electrophoresis of rat serum. This technique discloses whether the serum under analysis contains esterases (*i.e.* the enzyme hydrolyzing esters) and at the same time facilitates the separation of distinct molecular species of these enzymes.

Theory: The under lying principle can be readily understood, if a small piece of filter paper is soaked in serum and used as a sample. The gel gets moulded in the form of a flat thin plate and forms a part of an electric field. The sample is inserted near the cathodic end of the plate. Once the circuit is switched on, the charged molecules start moving in a direction and speed dependent on the charges, the amount of current passing and also on their size in relation to the pores of the gel. When electrophoretic run is completed, the molecules are separated and held up at different distances from the point of origin. Thereafter, these are rendered visible using the colouring reagent.

The advantage of this process is that at a time one plate can be used to analyze a number of samples, so that the differences in the esterase complement of the many sera may be detected under identical electrophoretic conditions. Differences encountered will reflect upon the genotype or species and age of individuals from the samples were obtained. This technique may be utilized for a wide range of comparisons.

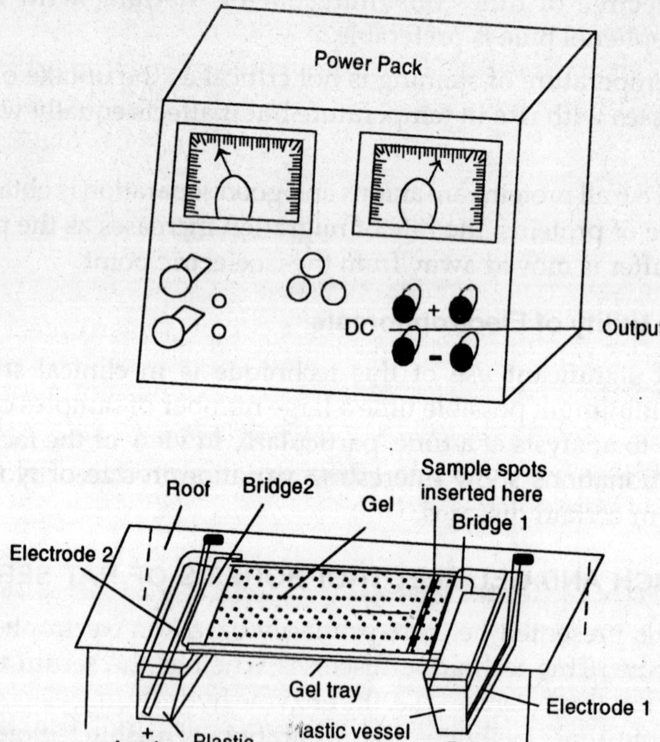

Fig. 7.4a. Assembly for horizontal starch gel electrophoresis of rat serum

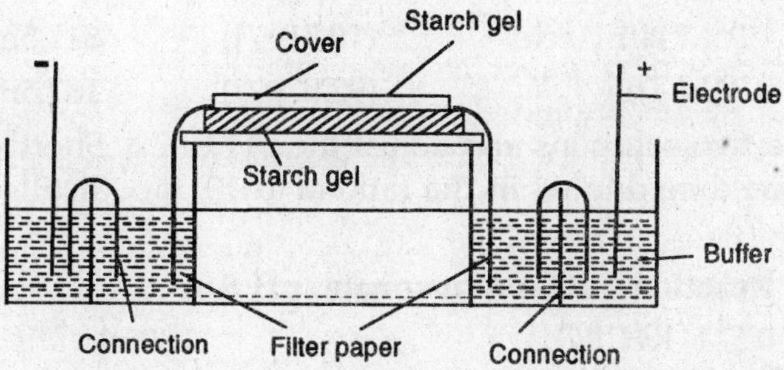

Fig. 7.4b. Diagramatic representation of horizontal starch gel electrophoresis

The apparatus

The diagrammatic illustration of the apparatus used in horizontal starch gel electrophoresis is quite simple as shown in Figures 7.4a and 7.4b. It consists of:

(1) **A Power Pack,** to maintain the supply of a continuous current at a constant voltage upto 300V and 150 mA.

(2) **Two plastic vessels,** slightly greater in dimensions than the tray and are filled with buffer solutions.

(3) **Bridges,** pieces of double folded lint linking the buffer sol with the gel on plate.

(4) **Gel Tray,** this is made up of perpex with 50mm rim in height all round. The gel is filled in this tray.

(5) **Roof:** This consists of glass plate covering the entire top of all vessels just for safety measures.

(6) **Safety microswitch** the circuit is closed when the roofing plate is pressed on it but on being lifted the circuit is interrupted.

Solutions, reagents and accessories

(i) **Starch:** Prepare the gel from hydrolyzed starch. Commercially it is available from BDH. In laboratory it is produced by hydrolysis of pure potato starch.

(ii) **Buffer**

(a) **Borate buffer for vessels**

Boric acid	18.6g
Sodium hydroxide	2.0g

Mix the two in a little quantity of water in a measuring flask and make up the volume to 1000ml by addition of more of distilled water.

(b) **Tris citrate pH 7.6 buffer for gel**

0.05M citric acid	(10.51g/l)	812.5ml
0.76M Tris	(92.07g/l)	187.5ml

Mix the two solutions and adjust the pH to 7.6. Shortly before use, it should be diluted in the ratio of 1: 10, mix distilled water (v/v).

(c) **Reaction buffer (phosphate pH 6.0)**

0.5M KH_2PO_4	239ml
0.5M Na_2HPO_4	0.26ml

Mix the two and raise the volume to 1000ml by adding distilled water and check pH.

(iii) Chemicals

(a) Non specific ester substrate: 1 naphthyl acetate

(b) Coupling dye: Fast Garnet GBC (Gurr's)

(c) **Reaction solution for 1000ml** This is prepared by dissolving 500-1000mg of Fast Garnet in 300ml of phosphate buffer (pH 6.0). Make up the volume to 100ml. Always prepare afresh before use and just before use dissolve 50-100mg of the substrate in a few millilitres of ethanol or methanol and add it to Fast Garnet solution.

(iv) Other accessories

(a) **Sample holders** Cut Whatman No. 1 filter paper into pieces 50 x 100mm or of any other convenient dimensions.

(b) **Fuse wire** finest wire, used for cutting through the gel plate.

(c) **Staining trays** of dimensions larger than the gel trays. These can be of plastic, enamel, glass or even aluminum trays or dishes. These should not be put in use for repeated number of times as they themselves get stained.

Procedure

For most satisfactory achievements proceed step by step as under:

1. Preparation of starch gel

Nearly 24g of starch in 200ml of buffer is required for use in a gel tray of approximately 20 x 12 x 0.5mm. For the preparation of starch gel, put 200ml of gel buffer (ii b) in 1000ml Perpex flask having a side arm for outlet, add 24g of starch and stir well until completely dissolved. Now, gently heat the flask with continuous stirring of solution just by shaking the flask. During this process the solution will gradually become viscous and opaque, but at 74°C it will again become more fluid and clear. As soon as this condition is arrived at, the flask should immediately be removed from flame. When bubble free gel solution has been prepared then promptly pour the gel into gel tray and ensure that the gel has spread uniformly and reaches above the rims. Allow it to stand for a few minutes and then gently cover the whole tray with glass plate. The gel plate will require 2-3 hours to set and it will be best if left overnight.

2. Application of samples

Lift the glass plate from the starch plate. At a distance of 4cm from the near side end of the plate make a trench by removing the ribbon of starch between two slits as directed leaving the trench empty.

With the help of a pair of forceps, take a piece of sample holder (filter paper piece) and dip it in the serum sample. Place it against face of one vertical wall of trench. Similarly place the other samples of holder, leaving a distance of 5mm or more between them.

Push the small portion of the starch gel towards the other, eliminating the trench. The plate is ready to be put in position bridging the two buffer vessels of the apparatus. Place it with the samples nearer the cathode.

Each bridge is a rectangular piece of lint wetted in the vessel buffer. No more than 3-4cms of the starch plate should be covered by the lint pieces on either side.

To avoid any sort of distortion, cover the surface of the starch with a piece of "Handi wrap", a very thin film that clings to the gel surface.

Safety precautions: Place the glass plate "roof" over the apparatus as shown in Figure 7.4 b, to prevent any one from inadvertently touching the circuit.

Switch on the current and regulate the voltage to about 10V per cm length of the gel. Make a note of the corresponding amperage (about 5mA per cm width of gel). The run should be made in a cool place. (The higher voltage speeds up the run but may cause overheating of the gels, which is not desirable).

3. Termination of the runs

The added advantage of this experiment is that a brown line moves in the starch. When this line has advanced 7-9cm from the sample holders, *i.e.* the place of origin, stop the run by switching off the apparatus simply by lifting the glass roof and then the gel tray. Pull the starch portion behind the sample holder and remove sample holder with a pair of forceps. The plate must now be sliced horizontally by means of fuse wire as directed. Discard the upper and lowermost sliced portions and use the middle one for staining. One can stain both slices by transferring them in a tray with reaction solution (iii.e.). These sliced pieces can be handled normally.

4. Staining of the gel slices

Submerge the sliced pieces with cut surface uppermost in the reaction mixture in a tray. The substrate, 1- naphthyl acetate as ester, is hydrolyzed by most of esterases. The enzyme release naphthol, which couples with the azo dye present (Fast Garnet) to produce a purple precipitate in the starch in the position whose various esterases have migrated during the run.

Within 10-20 minutes all bands and zones indicating the location of esterases will start appearing. Photograph the plate, if possible.

To avoid over staining, stop the reaction by replacing the solution with water. The plates can be kept for a long time under water in a cold room; otherwise at room temperature they soon became "fluffy" due to bacterial attack.

C. POLYACRYLAMIDE GEL ELECTROPHORESIS

Of many electrophoresis techniques, each based on use of different supporting medium, polyacrylamide gel electrophoresis (PAGE) is most common and widely used technique particularly for the separation and determination of molecular weights of proteins, nucleic acids, peptides, *etc.*

Superiority of PAGE over other types of electrophoresis

It has high degree of reproducibility and porosity, thus facilitating the separation of molecules having similar charges but with different shapes and sizes, especially suitable for resolving mixture of proteins. Using this technique, the macromolecules are readily separated on account of their minimum adsorption capacity, their lack of electroosmosis (movement of liquid under applied electric field through a fine tube), general suitability for *in situ* quantitative analysis and various types of histochemical analysis (the study of minute structures of the tissues of organism).

The Apparatus

It comprises of power pack capable of providing a stabilized current and controls for both voltage and current output. Power packs are available for low voltage use, which has an output of 0 to 500V and 0 to 150mA and can give either constant voltage or constant current. The electrophoresis unit (Figure 7.5) contains the electrode, buffer reservoirs, a support for the electrophoresis medium and a transparent insulating cover. Stainless steel electrodes may be used but some buffers cause corrosion, hence platinum electrodes are more satisfactory.

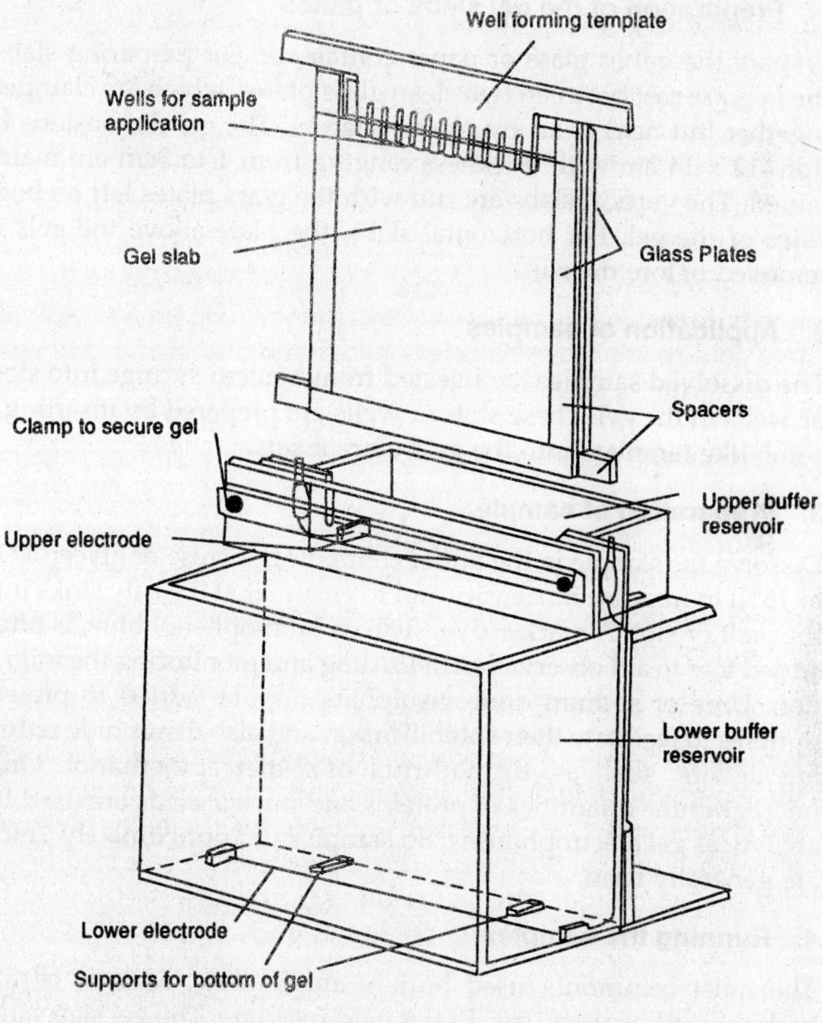

Fig. 7.5. Vertical gel electrophoresis unit

Although in this equipment gels can be run as horizontal slabs or vertical slabs but the vertical slabs are more frequent in use. Such units are commercially available for some analytical procedures. Slabs of gel carry more samples than cylinders (upto 25 samples, can be run on the single gel plate) making them more economical to use and enabling more samples to be compared with each other under identical conditions.

Procedure

For polyacrylamide gel electrophoresis (PAGE) proceed step by step as under:

1. Preparation of the gel slabs or plates

Prepare the gel in glass or paper containers. For preparing slabs, the gels are cast between two clean glass plates, which are clamped together but held apart by plastic spacers. The gel dimensions of upto 12 x 14 cm with thickness ranging from 1 to 3cm are maintained. The vertical slabs are run with the glass plates left on both sides of the gel. For horizontal slabs, the plate above the gels is removed before the run.

2. Application of samples

The dissolved samples are injected from a micro-syringe into slots or wells in the gel. These slots or wells are prepared by inserting a comb like template into the gel before it sets.

3. Preparation of sample

Dissolve the sample in the buffer containing sucrose or glycerol (10 to 15%) to increase its density and to ensure that it easily sinks into the well or slot. A marker dye, such as bromophenol blue, is often added to it to aid observation of loading and monitoring the migration. Urea or sodium dodecylsulphate may be added to protein samples to facilitate their solubilization and also disulphide reducing agents, such as, dithiothritol or 2-mercaptoethanol. Only microgramme quantities of proteins and nucleic acids are used for analytical gel electrophoresis. So samples of approximately 1mm^3 are generally used.

4. Running the samples

The most commonly used buffers are formate, acetate, citrate, barbitone, phosphate, tris, EDTA, and pyridine. The gel slab sandwiched between glass plates is placed in lower reservoir with the top of the gel in contact with the buffer in upper reservoir and lower part of gel also in contact with the buffer of lower reservoir, The gel thus completes the electrical circuit between the electrodes in the upper and lower compartments. Although the buffer surrounding the gel helps to disperse heat generated by the current, additional cooling may be required during long runs. This can be achieved by carrying out the run in a cold room or by circulating the buffer through a cooling system. The precise voltage and time required to obtain optimal separation will depend on the nature of the samples and the type of the gel used; but several hours at a few hundred volts are generally required. Marker dye, such as bromophenol blue for proteins and ethidium bromide for nucleic acids, enables the progress of the run to be monitored. After the run sample bands can be visualized.

5. Detection, recovery and estimation

The identification of components of unknown mixtures on the electrophoretogram after electrophoresis is achieved by comparing the migration characteristics of the resolved mixture with pure samples of known compounds under identical conditions to the test sample. If substances, naturally absorb or fluoresce under ultraviolet light and the supporting medium does not, this may be used as a means of detection. For example, nucleic acids and proteins can be identified by ultraviolet absorbance within the 260-280nm ranges in transparent polyacrylamide gel containing a low percentage of cross-linking agents.

Most biological molecules are colourless and have to be treated with reagents, which specifically produce stable coloured compounds.

Note

1. Excess dye or colour inducing reagent must be removed from the supporting medium by elution in a suitable solvent or by further electrophoresis. After the run is complete, the samples are fixed to avoid diffusion by soaking the gel in a fixing solution (*e.g.* 10% TCA) for 30 minutes. The samples can then be stained by immersing the gel in a suitable staining solution for 10 minutes. Several rinses in a de-staining solution (an aqueous solution containing 25% ethanol and 8% acetic acid) are required to remove excess stain.

2. Quantitative analysis is also possible for electrophoretogram in which the separated compounds are removed from the supporting medium. The most common method of recovery is to divide up the supporting medium into standard uniform lengths and to treat the supporting medium in a way that releases the test compounds.

Table 7.1: Staining reagents for visualising the resolved compounds of electrophoresis

Compound	Reagent	Comments
1. Proteins	Nigrosine in acetic acid or TCA	Very sensitive
	Bromophenol blue	Quantitative
	Lissamine green	Quantitative
	Coomassie Brilliant blue	Quantitative, very sensitive
2. Peptides	ClO_2 or NaOCl Chlorination followed by KI starch or benzidine aectic acid	Reacts with all NH_2 compounds
3. Nucleic acids	Methyl green pyronin	RNA red DNA blue
3. Polysaccharide	Iodine	Sensitive
4. Acid muco-poly saccharides	Toluidine blue in methanol water	Sensitive

Factors affecting electrophoresis: The following points are noteworthy and every care has to be taken to have most satisfactory results.

1. Electric Field

(a) **Voltage:** If the separation of the electrodes is d (distance) metres and potential differences between them is v volts, the potential gradient is v/d volts m^{-1}. An increase in the potential gradient will increase the rate of migration proportionally.

(b) **Current:** The rate of migration is proportional to both the current and the time.

(c) **Resistance:** Rate of migration is inversely proportional to the resistance of medium. It increases with the length of supporting medium and concentration of buffer.

2. The Sample

(a) **Charges:** The rate of migration increases with an increase in the net charge.

(b) **Size:** The rate of migration decreases for larger molecules due to the increased frictional and electrostatic forces.

(c) **Shape:** Molecules of similar size but different shapes, such as, fibrous and globular proteins, exhibit different migrations due to frictional and electrostatic forces.

3. The buffer

(a) **Composition:** The buffer should be such that it does not bind with compounds being separated as this may alter the rates of migration.

(b) **Concentration:** As the ionic strength of the buffer increases, the proportion of the current carried by the buffer will increase and the share of current carried by the sample will decrease, thus slowing its rate of migration.

(c) **pH:** The ionization of organic acids increases as pH increases whereas the reverse applies for organic bases; therefore their degree of migration will be pH dependent.

Precautions

(1) Electrophoresis unit should always be enclosed by a cover during use to minimize evaporation and to provide electrical insulation.

(2) Electrophoresis unit should be set at constant voltage and current because fluctuation of voltage causes heat effect which affects migration.

(3) The gel should be uniform otherwise it causes molecular sieving, which affects the migration.

8

Physiology of Blood

INTRODUCTION

The blood is a fluid connective tissue, which circulates throughout the body of a living animal. One cubic millimetre of blood of a normal human being contains approximately 50,00,000 cells.

The blood consists of formed elements suspended in liquid medium called the plasma. The formed elements are the red blood cells (RBC) or erythrocytes, the white blood cells (WBC) or leucocytes and the blood platelets. If a little quantity of an anticoagulant, such as oxalate, is added to a sample of blood, the clotting of blood is prevented. If the blood is allowed to stand in a test tube, the formed elements or blood cells settle down leaving at the top, a straw-coloured liquid, the serum.

Exercise No. 1

To study the formed components of the blood.

Theory: A quick method to study the composition of blood for the formed elements is to make a stained preparation of a thin uniform film of blood of a normal human being and its examination under high power of microscope for the presence, shape, size, colour, relative number and general appearance of different types of cells.

The normal blood of men contain about 50,00,000 RBC/cubic millimetre while the normal value of women is somewhat lower approximately to 45,00,000. The red cells contain no nuclei and are biconcave disc like in appearance.

In the blood of a normal person, white cells are present to the extent of 5,000 to 10,000 per cubic millimeter. There are several types of white blood cells, most of which are larger than red blood cells. All WBC's are nucleated.

Blood platelets are normally present to the extent of 3,50,000 per cubic millimetre. They are much smaller than red cells and are of irregular shape. They are devoid of nuclei and in a stained preparation they are visible in groups or clusters. Figure 8.1. shows the general appearance, relative size and other characteristics of different types of cells as revealed in the microscopic examination.

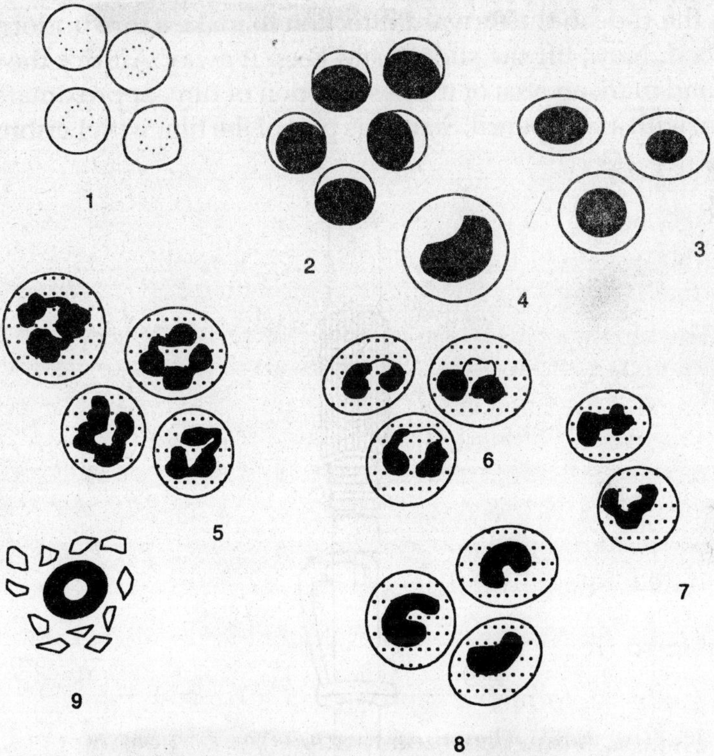

1. RBC
2. Small lymphocyte
3. Large lymphocyte
4. Monocyte or large mono nuclear
5. Polymorphonuclears (Neutrophils)
6. Eosinophils
7. Basophils (Mast cells)
8. Transitionals
9. RBC surrounded by blood platelets

Fig. 8.1. Various types of blood cells

Preparation of the stained blood film

Wash your hands and sterilize your one finger with rectified spirit. Either with the help of blood lancet (Figure 8.2.) or with a well-sterilized needle prick the distal end of your finger and gently press it. The blood will ooze out. For preparing a uniform blood film proceed step by step as shown in Figure 8.3. Just by touching transfer the drop of blood at one end of the clean and grease free slide A. Make a thin uniform film with the help of another slide B placed in front of the blood drop at 45°, drag it towards the blood drop. As soon as its lower end touches the blood drop, it will spread all along the lower border of the slide held at 45°. Now, push the slide in the forward direction to make a thin uniform film of blood. Now, lift the slide B and keep it away. Air dry the blood film and mark an area of the best portion of film, approximately 1½ sq.cm, with a wax pencil. Stain this part of the film with Leishmann's stain.

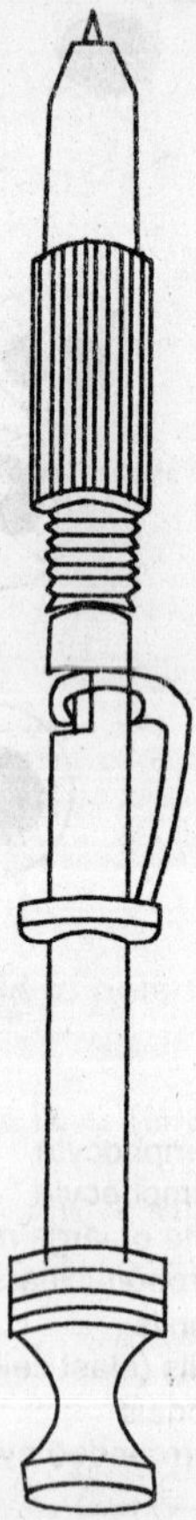

Fig. 8.2. Blood Lancet

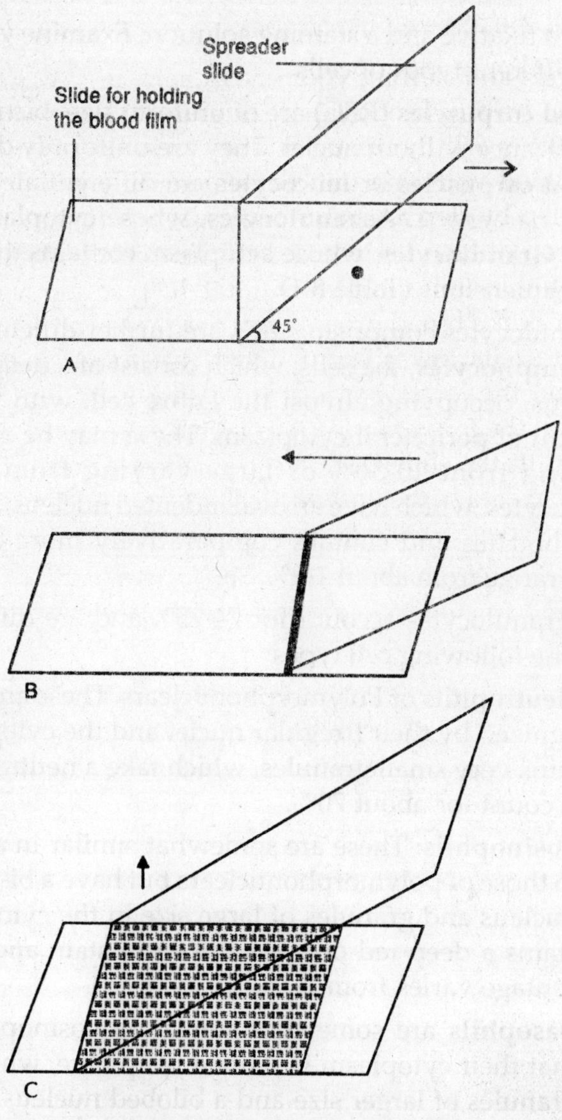

Fig. 8.3. Preparation of blood film

Staining the film with Leishmann's stain

Put a few drops of Leishmann's stain; say 2 drops, in the area marked with wax pencil and wait for 1-2 minutes. Do not allow the stain either to dry up or precipitate. After this add double the quantity of distilled water, *i.e.* 2-4drops. Bring about mixing the two by moving the slide sideways. Wait for 8-10 minutes. Now throw away the diluted stain and dry the film. Leishmann's stain acts both as a fixative and a staining solution. Examine your preparation for different types of cells.

Red blood corpuscles (RBC) are of uniform size, biconcave disc like in appearance without nuclei. They are uniformly distributed. **White blood carpuscles** or **leucocytes** are differentiated into two main types, namely: A. **Agranulocytes**, whose cytoplasm is non-granular. B. **Granulocytes**, whose cytoplasm contains the granules of varying dimensions (Table 8.1).

(A) Agranulocytes comprising 26% are further differentiated in: (a) **Lymphocytes**, the cells, which consist of a deeply stained nucleus, occupying almost the entire cell, with very small amount of peripheral cytoplasm. These may be either small varying from 20-30% or large varying from 2-6%; (b) **Monocytes**, which have an oval indented nucleus, larger than lymphocytes and contain comparatively more cytoplasm. They range from about 1-2%.

(B) The granulocytes account for 74-75% and are differentiated into the following cell types:

- (a) **Neutrophils** or Polymorphonuclears. These are easily recognized by their irregular nuclei and the cytoplasm contains very small granules, which take a neutral stain and account for about 70%.
- (b) **Eosinophils:** These are somewhat similar in appearance to those of polymorphonuclears but have a bi- or trilobed nucleus and granules of large size in the cytoplasm that stains a deep red or takes the eosin stain and their percentage varies from 1-2%.
- (c) **Basophils** are somewhat similar to eosinophils except that their cytoplasm is stained deep blue, which contain granules of larger size and a bilobed nucleus. Their percentage varies from 0.25-0.5%.

The Blood Platelets or **Thrombocytes:** These are small cells in the form of rounded or oval discs or spherical bodies measuring about one-third size of a mature red blood cell (RBC). There are approximately 3,00,000 of them in each cubic millimeter of blood. They play an important role in the control of bleeding after injury and in clotting of blood.

Exercise No. 2

Total count of blood cells.

Table 8.1: Shows the percentage of various types of leucocytes (WBC) in normal human blood.

	Type of cell	%Range
A. Agranulocytes	(a) Small lymphocytes	20-30%
	(b) Large lymphocytes (monocytes)	2-6%
B. Granulocytes	(a) Polymorphonuclear (Neutrophils)	65-70%
	(b) Eosinophils	1-2%
	(c) Basophils	0.25-0.5%
	(d) Transitionals	2-4%

Apparatus, reagents, *etc*.: Haemocytometer, Blood lancet, Micropipettes, Hayem's fluid, Turk's fluid, ether, alcohol, microscope, *etc*.

Haemocytometer: This instrument is meant for counting the blood corpuscles. Neubaur haemocytometer is most frequently used in laboratories. It consists of a double-cell slide with two sunken platforms. Each platform has one counting chamber. The two central platforms are bordered by an edge-shaped groove and when a special cover glass is kept in position, a depth of 1/10mm is maintained over the markings (Figure 8.4).

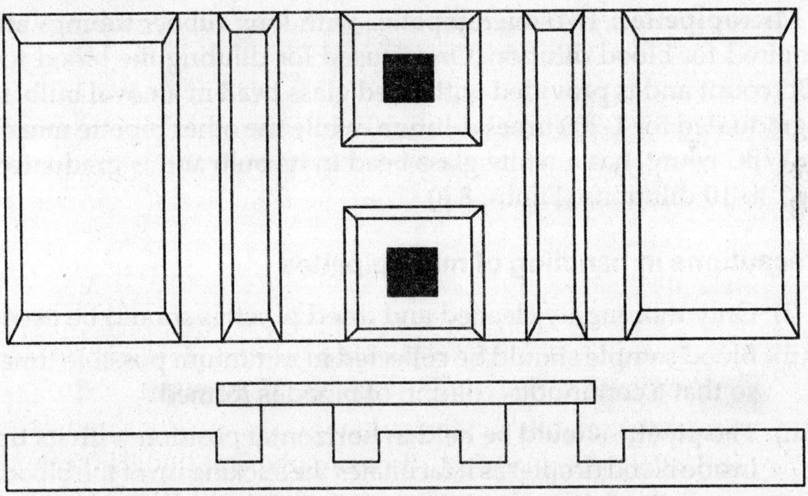

Fig. 8.4. Neubauer counting chamber

Each counting chamber is formed by several straight perpendicular and horizontal lines, enclosing squares of various measurements. The biggest square is 3x3mm with an area of 9sq. mm. It is divisible in 9 squares of 1x1mm each. The central square is further divided into 25 small squares, each square measuring 1/5 x 1/5 = 1/25 sq. mm. Each of these squares is further divided into 16 smaller squares, so that there are 400 smallest squares in all, each with an area of 1/400sq. mm (Figure 8.5).

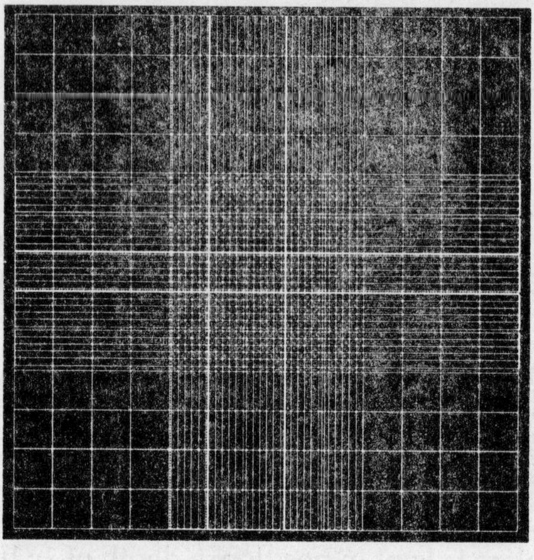

Fig. 8.5. Neubauer ruling on a blood-counting chamber

Micropipettes: Two micropipettes with long rubber tubings are required for blood dilution. One is used for diluting the blood for RBC count and is provided with a red glass bead in an oval bulb. It is graduated for 1-200 times dilution, while the other pipette meant for WBC count, has a white glass bead in its bulb and is graduated for 1 to 10 dilutions (Figure 8.6).

Precautions in handling of micropipettes

(i) Only thoroughly cleaned and dried pipettes should be used.

(ii) Blood sample should be collected in minimum possible time, so that a continuous column of blood is formed.

(iii) The pipette should be held in horizontal position with its tip inside blood droplet as it facilitates the sucking up of the blood.

The accuracy and success in this exercise and in handling the pipette, all depends upon the practice and experience.

1. Procedure for counting RBC

Significance of RBC Count

One of the most common tests in a pathological laboratory is the determination of the number of various formed elements present in the blood. The reason for this test is that there is a wide fluctuation in total number from the normal condition to the diseased one, e.g. in anemia the RBC counts are low and in many infections, leucocyte counts are high.

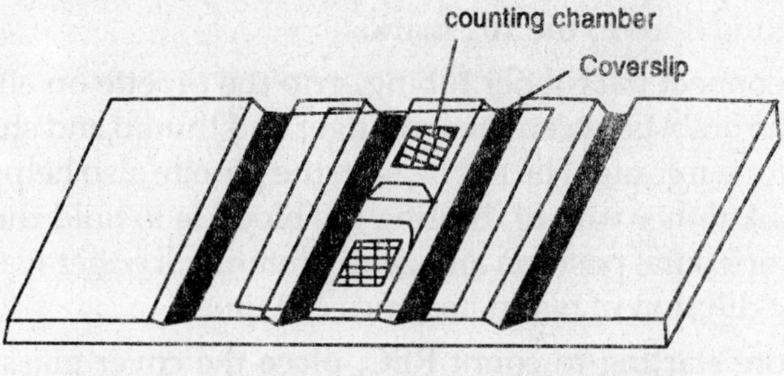

Glass slide with counting chambers (Neubaur Haemocytometer)

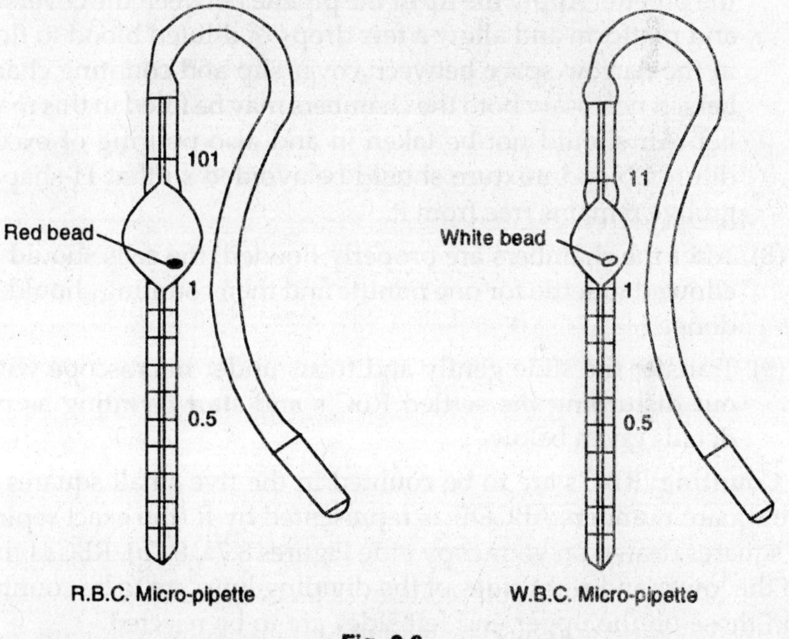

Fig. 8.6

For more accurate results proceed step by step as under:

(1) Clean and dry the graduated diluting micropipette (RBC pipette dilution, 1:200) provided with red glass bead.

(2) Sterilize your middle finger or of your partner with a small cotton swab or a pad dipped in 90% alcohol.

(3) Prick the middle finger with the blood lancet and wait till the blood flows freely. Do not squeeze the finger.

(4) Wipe away the first few drops and then slowly suck the blood in the pipette upto 0.5mark. Care should be taken to avoid the entry of air, as it will result in air bubble formation. The excess blood may be wiped off at the end of pipette on filter paper.

(5) Immediately mix the blood with Hayem's diluting fluid, by sucking it up to the 101 mark.

(6) Disconnect the rubber tubing, grip the pipette on either side of the bulb, between the forefinger and thumb and shake gently for a minute. The red bead in the pipette also helps in mixing. Another way of diluting the blood is to hold the pipette in horizontal position and gently rotating it either way. In this way dilution of blood becomes 200 times.

(7) Before starting to count RBC, place the cover glass/slip on the haemocytometer. First reject 3-4 drops of the mixture from the pipette. Apply the tip of the pipette between the coverslip and platform and allow a few drops of diluted blood to flow in the narrow space between cover slip and counting chambers. If necessary both the chambers may be filled in this manner. Air should not be taken in and also pouring of excess diluted blood mixture should be avoided so that H-shaped groove remains free from it.

(8) After the chambers are properly flooded, the cells should be allowed to settle for one minute and then counting should be done.

(9) Transfer the slide gently and focus under microscope without disturbing the settled RBC's and start counting as per details given below.

Counting: RBC's are to be counted in the five small squares in the square numbers ABCDE as represented by R (the exact replica of squares drawn on your copy vide Figures 8.7a, 8.7b). RBCs lying on the lower and right sides of the dividing lines are to be counted and those on the upper and left sides are to be rejected.

Calculation

Number of RBCs per cubic mm

$$= \frac{\text{Number of cells counted} \times \text{Dilution} \times 4000}{\text{Number of small squares counted}}$$

Suppose total RBCs in 5 smaller squares are: A+B+C+D+E

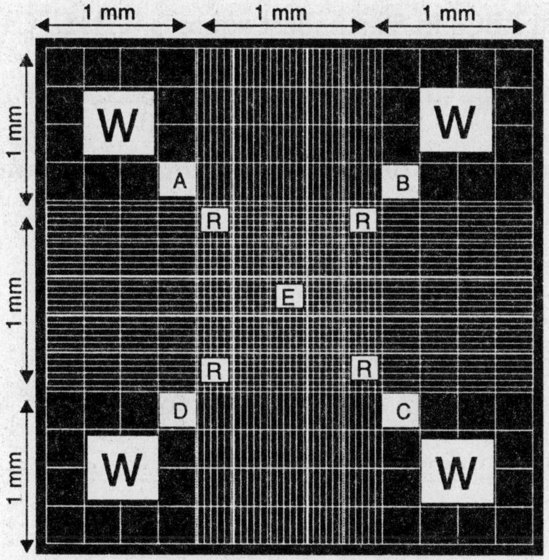

Ruling of the improved Neubaur slide
R denotes small squares for red cell count
W denotes big squares for white cell count

Fig. 8.7a. Counting platform

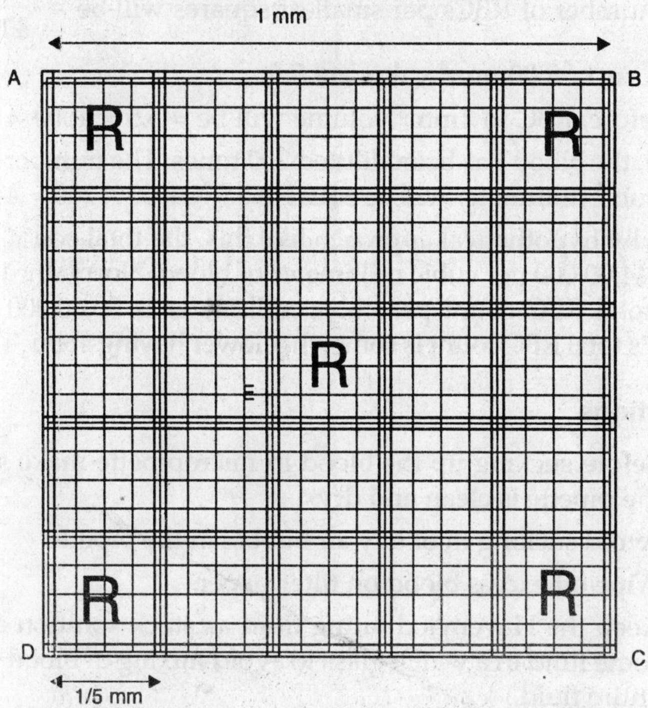

Fig. 8.7b. Magnified view of the central ABCDE squares

Therefore total RBCs in 80 smallest squares = A+B+C+D+E

One smallest square will contain $= \frac{A+B+C+D+E}{80}$

Therefore total number of RBCs in 400 smallest squares

$$= \frac{A+B+C+D+E \times 400}{80}$$

But height in chamber = 0.1mm

Dilution of blood = 200 times

Therefore, one cubic mm of blood will contain

$$= \frac{A+B+C+D+E \times 400 \times 10 \times 200}{80}$$

$$= A+B+C+D+E \times 10{,}000 \text{ RBCs.}$$

This can be understood in another way:

Suppose the number of RBCs in 5 smaller squares or 80 smallest squares is (84+86+88+90+94) = 442

The number of RBCs per smallest squares will be $= \frac{442}{80} = 5.5$

i.e. RBC in 1/4000 mm^3 value are 5.5

Therefore, RBC in 1mm^3 volume will be = 5.5 × 4000 = 22,000

Since, the blood has been diluted 200 times. The number of RBCs in one cubic millimeter will be equal to 5.5 X 4000 X 200 = 44,00,000.

Thus by hypothetical approach like this, the total count of RBCs will be 44,00,000 per cubic millimeter of blood. Normally a healthy man's total RBC count per cubic millimeter is 50,00,000 while a women's total RBC count is somewhat lower having about 45,00,000.

Precautions

(1) Before sucking up the blood in micropipette make sure that the pipette is clean and dry.

(2) Avoid sucking in of any air bubble in the pipette.

(3) Wipe off excess blood on filter paper.

(4) Keep the Hayem's diluting fluid as stock solution and take some fluid in a watch glass to avoid mixing of blood with the entire fluid.

(5) Avoid filling the H-shaped groves of the haemocytometer.

(6) Thorough mixing of blood and diluting fluid should be ensured for uniform distribution of RBC. No air bubble should enter the counting chamber.

Hayem's fluid consists of

Mercuric chloride	0.5g
Sodium sulphate	5.0g
Sodium chloride	1.0g
Water	200ml

2. Procedure for total WBC or Leucocyte Count

Significance: A leucocyte count is of great help to a physician in the diagnosis of disease, like appendicitis, where in acute cases the leucocyte count often reaches 15,000-20,000. The disease in which the leucocyte count increases to the greatest extent is leukemia in which the total counts to the extent of 20,000 to 5,00,000 are reported. This disorder of blood is referred to as *leucocytosis.*

The leucocytes are counted in much the same manner as the RBCs except that the diluting fluid is 1.5 percent acetic acid or Turk's fluid, which destroys the RBC's and make the leucocytes more readily and easily visible. Because of the smaller number of WBCs, the dilution is 1-20 instead of 1-200 as in the case of RBC count. The count is made under the low power of the microscope and an area of 1sq. mm is the unit counted.

Procedure

(1) For perfect assessment proceed step by step as under:

(2) Clean and dry the WBC micropipette.

(3) Sterilize your middle finger or your partner's middle finger with a small cotton swab soaked in 90% alcohol.

(4) Prick the middle finger with the blood lancet.

(5) Discard first few drops, then suck up the blood upto 0.5mark on micropipette and immediately dilute it with diluting fluid (1.5 percent acetic acid or Turk's fluid)

(6) After soaking excessive fluid on the filter paper, rotate the horizontally held micropipette sideways so as to allow the uniform mixing of the blood with the diluting fluid. The blood is diluted 20 times in WBC pipette with any diluting fluid.

(7) Place a clean cover slip on the counting chamber and charge

the slide with the mixture after discarding the clear fluid in the capillary part of micropipette. Allow the corpuscles to settle down and make a count under the low power of microscope as detailed below.

Counting: The leucocytes are recognized under low magnification of the microscope by the refractile appearance and by slight colour given to them by stain contained in the diluting fluid (in case of Turk's fluid). The number of WBC present in four chambers of the corner of 1sq.mm in the central ruled area on the sides of the counting chamber of haemocytometer (denoted by w in the replica of squares drawn in Figure 8.7a, b), *i.e.* the counting is to be continued in the outer four squares forming WBC area. The WBCs lying on the lower and right sides are to be counted whereas those present on the upper and left sides is discarded

Calculation

Number of WBCs per cubic mm

$$= \frac{\text{Number of cells counted} \times \text{Dilution} \times 10}{\text{Number of squares of 1 sq. mm counted}}$$

Since dilution is 20 times and cubic capacity of each area counted is 1/10 cubic mm, the volume is 1/200 cubic mm.

Say e.g. WBC counted in 4 squares is = x

WBC in 1/200 cubic mm = x

WBC in 1 cubic mm = x × 200 × ¼

= x × 50

Another simplified way to understand the same is:

Let us presume that the number of WBC in 4 outer squares is

= (9+13+10+13) = 45

That means that the number of WBC in 1/200 cubic mm = 45

Therefore, the number of WBC in 1 cubic mm = 45 × 200× ½

= 4500 /cubic mm.

For favour of information of general interest a student must know that the total WBC count of a normal healthy man varies from 4000 to 6000 WBC per cubic mm of blood. In case there is marked fluctuation, that means either there is some flaw in manipulation or counting or calculations.

Precautions

Same as in case of RBC count except that in case of WBC count, the use of freshly prepared diluting fluid gives more accurate results.

Turk's fluid consists of Glacial acetic acid	1ml
1% aqueous solution of Gentian violet	1ml
Distilled water	1000ml

Exercise No. 3

Differential leucocyte count (DLC) in human blood

Requirements: Clean slides, Leishmann's stain or Wright's stain (which contain eosin, a red dye, methylene blue, a blue one), sterilized needle, microscope, cotton, alcohol, distilled water, immersion oil.

Significance: The proportion of different types of WBC varies from the normal values according to the nature of disease. Not only this, sometimes other types of leucocytes may also make their appearance.

In case of *appendicitis* the percentage of polymorphonuclears may vary from 80-90 percent or even more. This is indicative of an immediate surgical operation of appendix. In case of lymphatic leukemia, the number of leucocytes is greatly increased. In infection of mononucleosis, the percentage of large monocytes is also increased. Temporary presence of intestinal parasites often results in an increase in the percentage of eosinophils as in case of rheumatoid arthritis. Eosinphil count forms a good basis for the differentiation between rheumatoid and other types of arthritis.

Procedure

Carefully proceed step by step to achieve satisfactory results.

(1) Sterilize the middle finger with an alcohol (90%) soaked cotton swab.

(2) Prick the finger with a sterilized needle or blood lancet, wipe away the first one or two drops of blood.

(3) Take the next drop on a clean and dry slide at one end, place slide on a smooth neat and clean surface holding it steadily with left hand. With the help of a second slide (spreader slide) held at 45° draw a thin uniform film of the blood. Remove the spreader slide and air-dry the film.

(4) Mark an area of uniform film with a wax pencil and stain with a few drops of either Leishmann's stain or Wright's stain, whichever is available. Wait for 1minute and then carefully add distilled water to the stain. The quantity of distilled water should be double that of stain used. Mix the water and stain well by sucking the water in and out of the pipette. After sometime a greenish metallic scum will appear on the surface of mixture. Allow the diluted stain to act for 7-10 minutes. Thereafter drain off the stain and wash the film for 10-15seconds with distilled water. The colour of the film will rise to pink.

(5) If the colour is deep purple then again wash the film with distilled water.

(6) Dry the slide and examine under the low magnification of microscope for DLC as under:

(a) First examine the stained preparation to get an idea of different types of leucocytes under low power.

(b) Then examine under high power or oil immersion lens/ objective.

(c) In a longitudinal strip of the smear, count the various kinds of WBCs, maintaining a record on paper of each type of WBCs and calculate the percentage as under:

Calculation

$$\text{Percentage of different types of WBCs} = \frac{\text{No. of a type WBC} \times 100}{\text{Total No. of WBC}}$$

Precautions

(1) While drawing the blood film, make sure that the blood is spread evenly with no clogging.

(2) The film should always be air-dried but never to be dried over flame.

Exercise No. 4

To determine the blood group of ones own blood.

Materials: Test sera A and B, micro-slides, micro-dropper, lancet and glass marking pencil.

Theory: Exhaustive studies of human blood have shown that it differs from person to person on the basis of certain proteins in the plasma membrane of RBC's and coded by genes. These are of two types namely "A" and "B" and are called antigens (=agglutinogens-chemically glycoproteins). Two antibodies (=agglutinins), which can react with these antigens, occur in the blood plasma. These have been designated as anti-A or 'a' and anti-B or 'b'. The antibody 'a' reacts with antigen A, and the antibody 'b' with antigen B, thus bringing clumping together of blood cells. On the basis of antigens and antibodies reactions, four major types of the blood groups have been recognized in human beings. These include A, B, AB, and O. The group A is characterized by having antigen A and antibody b; group B by having antigen B and antibody a; group AB by having both antigens A and B but no antibodies and group O by having both antibodies and no antigens. The blood groups were first described by K. Landsteiner in 1900, who also formed a law which states that (i.) if an antigen is present on RBC, the corresponding antibody must be absent from the plasma, and (ii) if the antigen is absent in RBC the corresponding antibody must be present in the plasma. In addition to these, several other types of blood groups, namely the Rh, MNS, P, Lutheran, Kell, Lewis, Duff, Kidd, *etc.* have also been identified. But A, B, AB and O blood groups are the most commonly and universally found and accepted.

In human blood, two types of antigens, namely A and B, are found associated with RBCs and two types of antibodies, namely a (anti A) and b (anti B) exist in plasma. Agglutination is caused due to antigens and antibodies reactions. It has been found that a person of blood group A has antigen A on RBC and antibody b in plasma. Similarly, a person of blood group B, has an antigen B on RBC and antibody a in plasma. A person with group AB, has antigen A & B on RBC but no antibodies in plasma. Persons of group O have no antigen on RBC but have antibodies a and b in plasma (Table 8.1).

Table 8.2. Showing compatible blood groups for transfusion

Blood Group	Antigen on RBC	Antibodies in plasma	Can donate to	Can receive from
A	A	b (anti B)	A, AB	O, A
B	B	a(anti A)	B, AB	O, B
AB	A&B	Nil	AB	A,B,AB,O
O	Nil	a & b	A,B,AB,O	O

In case of serious loss of blood due to injuries and diseases, the blood is transfused in order to restore normalcy. But, before transfusing the blood it becomes mandatory to find out the blood group of the person to whom the blood is transfused. A person who receives blood is called the recipient and other who donates as the donor. The most important consideration in blood transfusion is the reaction between the antigens of the donor RBC and antibodies a person might have in RBC cells. A serum called combine natural antibodies capable of reacting with those antigens not present in its own blood should not be agglutinated or haemolyzed by the serum of recipient and the vice-versa. Individuals having O group are *universal donors,* since their blood can be administered to all individuals because of the lack of antigens on such RBCs. Persons belonging to AB group are termed as *universal recipients,* since they can receive blood from a person of any group. This is because they do not have antibodies in plasma to react against antigen of donor's blood. A person of group A can receive blood from O & A, of group B from O& B, of group AB from A, B, AB and of group 'O' from O only (Figure 8.8 and 8.9 and Table 8.2).

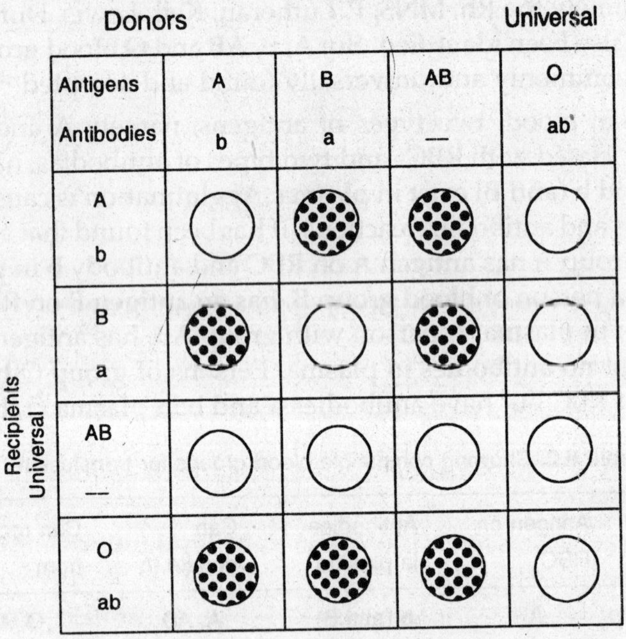

Fig. 8.8. Compatible blood groups for transfusion

Procedure to determine blood groups

(1) Diluted stock solution of antisera of group A and B should be

prepared in advance or procured from market and stored at low temperature.

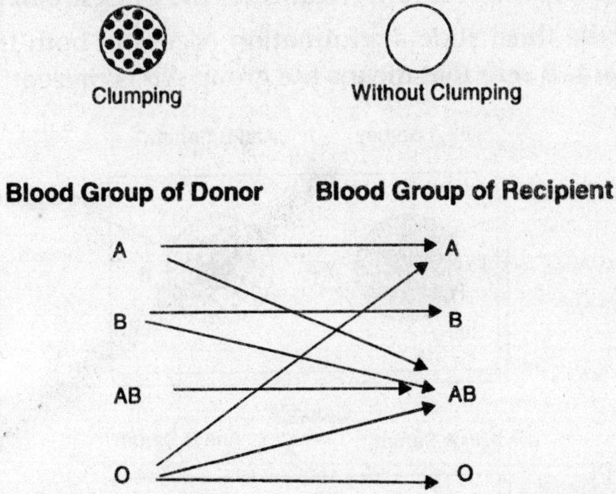

Fig. 8.9. Representation of the compatibility of blood groups in transfusion.

(2) Take 4 cleaned and dry slides labelled as No. 1,2,3 & 4 and place them over the white paper sheet already placed on a clean horizontal surface. In each slide mark two oval areas with a wax pencil one at left end marked A, another on right side marked B. On each slide, in area marked A, place one drop of anti A serum and on area marked B, a drop of anti B serum.

(3) In a test tube take 4ml of normal saline having a few drops of 3.7% sodium citrate then with the help of a lancet prick your own sterilized finger and mix a few drops of your blood in the test tube and gently shake to bring about uniform mixing of all the contents in the test tube.

(4) Place one or two drops of the mixed blood from the test tube in each A and B marked areas or a drop of blood directly from the person whose blood has to be tested in each area one by one in each slide.

(5) With the help of separate but fresh toothpicks, bring about the mixing of blood sample and serum in each marked area for 2-3 minutes in both the areas A and B of each slide and observe the reaction as under:

Observation

(1) In the first slide if agglutination occurs in anti A serum but

not in anti B serum it indicates the presence of group A.

(2) On the second slide if agglutination occurs in anti B serum but not in anti A serum it indicates the presence of group B.

(3) If in the third slide agglutination occurs in both the anti A and anti B sera that means the group AB is present.

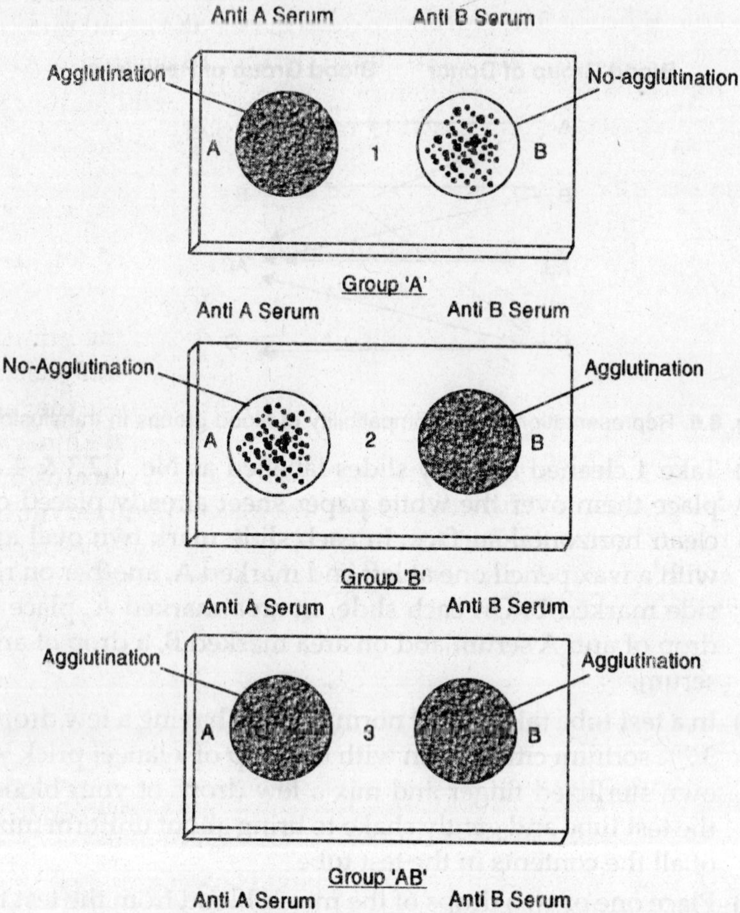

Fig. 8.10. Diagrams illustrating the procedure involved in the determination of blood group of a human being

(4) If in the fourth slide no agglutination takes place in both the anti A and anti B sera that means the group O is present.

Result: Report the result as concluded in four observations and confirm with illustration given in figure 8.10.

When blood of a person is taken out and allowed to clot, the residual liquid plasma part is called the serum. It contains all plasma antibodies of that person. To determine ABO blood groups, blood area-containing antibodies of that person. To determine ABO blood groups, blood area containing antibodies a and b, respectively, are prepared or purchased from the market.

Exercise No. 5

Determination of Rh factor in human being.

Materials: Clean slide, anti-Rh serum.

Theory: Besides the blood groups discussed above, the human blood may also contain another factor which is quite independent of the blood groups. This factor has enabled to differentiate the human blood further into two groups namely Rh-positive and Rh-negative. This factor is of great significance during blood transfusion. Rh stands for rhesus. When the rabbits are injected with erythrocytes of rhesus monkey called *Maraceous rhesus,* an anti-rhesus serum (or anti Rh) is developed. This anti-rhesus serum on being tested brings about the agglutination of the erythrocytes of monkey as well as a great percentage of human population. About 97% Indians possess Rh positive factor and the rest Rh negative factor.

Procedure

(1) Take a neat and clean air-dried microslide and mark an oval area with the help of a wax pencil in the centre of the slide.

(2) Prick your sterilized finger with a blood lancet and after discarding the first one or two drops, place the drop of blood in the marked area.

(3) Add a drop of anti rhesus serum (anti Rh procured from market) to it and mix it gently with a toothpick.

(4) Gently warm it over a low flame of burner avoiding overheating.

Observation and Result

If the agglutination takes place then the person concerned with the blood sample is called Rh positive (Rh+) otherwise Rh negative (Rh-).

Exercise No. 6

Determination of the coagulation or clotting time of the blood of a human being.

Materials: Capillary tubings of 1.5mm internal diameter, blood lancet or sterilized needles, cotton and filter paper, rectified spirit (90% ethanol).

Theory: If the blood of a human is withdrawn and put into a vessel it coagulates, or clots (changes into red gelatinous mass). With the lapse of time this clot slowly contracts leaving a yellowish fluid, called the serum, in approximately 10 minutes. This innate property of blood to coagulate is essential to life. In the absence of this property, there is no self-sealing device and results in haemorrhages (continuous bleeding).

Procedure No. 1

There are a number of methods for the determination of clotting time. Two such are given here as under:

(1) Sterilize your finger with a cotton swab soaked in 90% alcohol.

(2) Prick your finger with a blood lancet and discard one or two drops of blood oozing out at the onset of pricking.

(3) Take thoroughly cleaned capillary tubing and fill it with fresh drops of oozing blood.

(4) Carefully break short sections of uniform length (approx. 1cm) of the tube at regular intervals, say ½ minute. In breaking the sections of the tube a small file should be used. Gently separate the two ends by breaking the sections of the tube already removed. Go on repeating this process till the formation of the threads of fibrin spanning the gap between the two broken ends of the tubes takes place. Note after how many intervals of 1 minute this has happened. This will correspond to the clotting or coagulation time of the blood of the person whose blood was examined. This clotting time may differ from person to person but in case of the same person it will show a consistency.

Result: By simple calculation report the clotting time.

Procedure No. 2

(1) Prick your sterilized finger with blood lancet.

(2) Transfer one or two drops of freshly oozing blood on a neat and clean slide and immediately switch on the stopwatch.

(3) Thereafter, after every 30seconds drag a clean needle slowly through the blood droplet.

(4) Go on repeating it and switch off the stopwatch, till the moment you see that a fine thread is being pulled up by the needle.

(5) This is the moment when blood has started coagulating.

(6) Work out the actual time taken by the blood to undergo coagulation.

Result: Normal clotting time in minutes.

The normal clotting/coagulation time of man's blood is from 2-5 minutes.

Exercise No. 7

To determine the bleeding time of a normal human being.

Theory: It is actually the time required for a small cut or injury to stop further bleeding. This bleeding time may or may not correspond with the coagulation time of blood because it depends on (a) on the efficacy of the tissue fluid in accelerating the coagulation, (b) on the elastic nature of the skin, and (c) the mechanical and chemical action of the thrombocytes. On an average, the bleeding time varies from 1minute to 3 minutes but in certain cases it may be prolonged upto 7-8 minutes or sometimes even more, still further much more as in haemophilia.

Procedure No. 1

(1) Sterilize the tip of your finger.

(2) Prick the finger with a blood lancet.

(3) Gently go on pressing your finger with uniform pressure at interval of 30 seconds, till the oozing of blood completely ceases.

(4) During this period, at each interval of 30 seconds, go on soaking the blood with a blotting paper or filter paper strip till you reach a stage when no blood spot is formed. The total number of strips bearing blood spots 1 x 30 will be the bleed-

ing time, As for instance, let us say there were six blood spot bearing strips, each after 30seconds. That means there are 5intervals of 30seconds. Therefore the total duration will be 5 x 30 = 150 seconds.

Therefore, the bleeding time will be $\frac{150}{60}$ = 2.5 minutes

In each successive strip, the size of the blood spot will go on decreasing and a time will come when no more spot will be formed, *i.e.* the blood will completely cease or stop to ooze out.

Procedure No. 2

(1) Prick your sterilized finger with blood lancet.
(2) Collect the blood in a clean test tube.
(3) Switch on the stopwatch.
(4) Place the test tube immersed in a water bath set at 37°C.
(5) Check the blood flow in the test tube after every 30 seconds by tilting it.
(6) Immediately switch off the stopwatch when the blood flow stops in the test tube.
(7) Take three or four such readings with fresh punctures of one and the same person.
(8) Work out the average time to minimize the experimental error.

Exercise No. 8

To measure the blood pressure of man.

Materials: Sphygmomanometer, Stethoscope

Theory: The blood pressure is defined as the pressure, which is exerted by the blood on the walls of blood vessels. Usually the flow of a liquid is from a place of higher pressure to that at lower pressure. The circulation of blood indicates that it is on account of variation in pressure levels, which is persistently highest in arteries as compared to the lowest in veins; else the circulation of blood may not be possible.

The contraction of the ventricle is the main cause for the development of blood pressure. During the peak period of ventricular contraction, pressure in arteries reaches a maximum. This is the systolic pressure. During its relaxation the blood pressure falls down. It reaches a minimum just before beginning of the next cycle. This minimum is called the diastolic pressure

In human beings, the principle involved in the measurement of blood pressure is to apply pressure on an artery in the upper arm, while the observations of the pulse are perceived below the compressor. Artificially the pressure is applied by pumping air into a rubber bag contained in an armlet strapped above the elbow. The bag is connected to manometer, which measures the pressure of the air. The pressure just sufficient to obliterate the pulse, measures maximum pressure in the artery, the so called systolic pressure.

The instrument

The instrument used in the measurement of blood pressure is commonly known as sphygmomanometer. The two types of sphygmomanometer namely the aneroid type and the other one called mercury column type (Figure 8.11) are more frequently used. Each of these consists of an airtight pressure sleeve with a rubber bulb connected by rubber tubing to the sleeve and through which by pumping the air pressure sleeve is inflated. In both types a fine adjustment screw valve is also provided with the bulb for gradual deflation. The difference between the two lies in the dial.

Fig. 8.11. Mercurial B.P. Apparatus

The aneroid type of sphygmomanometer is provided with a graduated circular dial and a centrally located needle, whereas the mercury type has a mercury column graduated 0-300mmHg. A pressure tube from the pressure sleeve in case of aneroid type opens into dial box, while in case of mercury type; it opens into a mercury reservoir. Depending upon the blood pressure, the mercury is pushed into the vertical column and the pressure readings are made out from the graduated scale.

Procedure

The method described here enables us to measure both the systolic and the diastolic pressures at a time. Ask your partner or any other person, who's BP is to be checked to sit at ease on a stool or chair. Roll up the shirtsleeves on the upper left arm. Now, place the airtight rubber pressure sleeve snugly around the arm without much pressure and wrap the remaining strip around the arm and tie it under pressure sleep or with latest gripper device. Now following Ausculatory type method, measure both the systolic and diastolic pressures as per procedural details given below:

Place the bifurcated arms of a stethoscope in the ears and the diaphragm over the brachial artery close to the pressure sleeve as shown in Figure 8.12 and start pumping the air into pressure sleeve

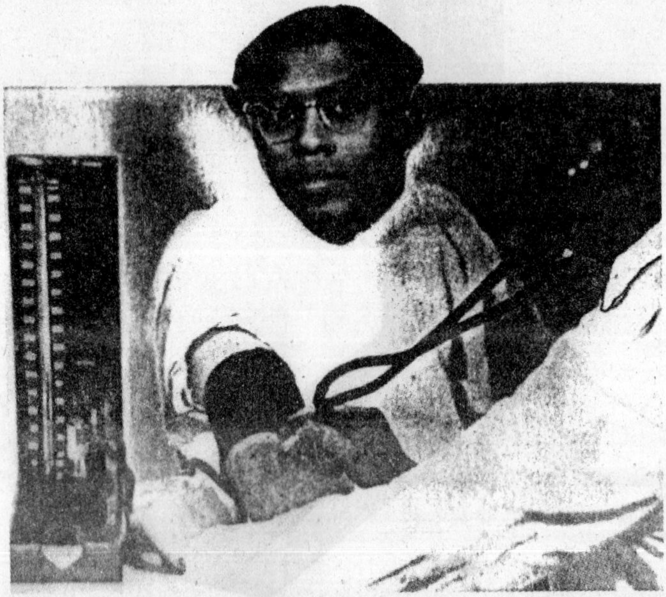

Fig. 8.12. Showing a diagram/photograph of a person undergoing blood pressure measurement

by pumping alternatingly the rubber bulb, till the mercury column reaches to a height of 150-200mmHg or dial needle at 150-200mmHg and by using the pressure release valve gradually go on deflating the pressure sleeve. Have a constant eye on the graduated scale or dial of sphygmomanometer and be vigilant for the point when the sound of the pulse appears in the ears. At a certain point a sound is heard with each heart beat, this indicates that the pressure in the brachial artery, which is capable of overpowering the outside pressure that means the flow of blood through the artery has started. Now again increase the pressure gradually and start deflating the pressure sleeve and note the highest pressure point at which the heartbeat is heard. This is the systolic pressure and is equivalent to the high blood pressure, which is usually 120mm Hg in normal healthy person. For ascertaining the diastolic pressure, gradually continue to deflate further the pressure sleeves and notice the change from sharp and clear audible clicking to murmuring sound of the blood flow and finally its disappearance at a particular point. This is diastolic pressure or low blood pressure point, usually 80mm Hg in a healthy normal human being. Repeat the process twice or thrice, compare the readings of high and low BP values with each other so as to be doubly sure about the final verdict. While measuring the BP also maintain a record of pulse rate.

The difference in pressure between the systole and diastole is called the pulse pressure and is normally 30-50mm Hg. The lower limit of systolic pressure in normal adult is estimated at approximately 105mm Hg and upper limit of systolic pressure at 150. In women the blood pressure is from 5-10mm Hg lower than in men.

Table 8.3: Normal blood pressure range (in mm of Hg)

S. No.	Human stage	Blood pressure	
		Diastolic	Systolic
1.	In infancy	50	70-90
2.	In childhood	60	80-100
3.	During the adolescent period	60	90-100
4.	In the young adult	60-70	110-125
5.	As age advances it increases	80-90	130-150

Factors responsible for maintaining BP

(1) The pumping force of the heart.

(2) The quantity of circulating blood.
(3) The viscosity of blood.
(4) The elasticity of the walls of blood vessels.
(5) The peripheral resistance.
(6) The mental equilibrium.

Exercise No. 9

Determination of the pulse rate of a man.

Materials: A timepiece or a wrist watch.

Theory: The so-called pulse is caused by the systolic activity of ventricle, which sends spurts of blood into the aorta. Due to this, the elastic wall of the aorta at the site of thrust of blood distends and contracts. This results in the transmission of the wave of alternate stretching and contraction of all the arteries as well as the arterioles. This very phenomenon is called as the pulse wave or the pulse. Each pulse or pulse wave thus comprises of a single stretching and contraction.

Table 8.4: Pulse rate in adult man and woman and at different ages.

Adult male	70-72/min (range 70-80)
Adult female	78-82/min (range 78-85)
Foetus	140-150/min
New born	130-140/min
First year	115-130/min
Second year	110-115/min
Third year	95-100/min
Seventh year- Fourteenth year	80-90/min
Over fifteen year	70-80/min
Old age	75-80/min

Procedure

Ask a person to relax and then place your finger over any artery lying close to the skin, preferably over the radial artery near the wrist. This is the usual and ideal method, which is generally used by the medical practitioners. When you start realising the pulse, count the number of pulses per minute and also visualize if there is any fluctuation in the regularity of the pulse. It should be noted here that the pulse is felt not due to the actual arrival of the blood at the wrist where pulse is usually felt, but due to the transmission of the wave of contraction and relaxation that have initially originated due to blood suddenly entering the aorta along the elastic walls of the arteries.

The number of pulses per minute will be the pulse rate. The pulse rate usually varies from 70 to 72 (range 70-80) in men and 78-82 (range 78-85) in women. However, the pulse rate shows variation in accordance with age (Table 8.4).

Exercise No. 10

Determination of factors affecting the blood pressure of man

Procedure

The influence of various factors on blood pressure can be worked out by using either type of sphygmomanometer, described earlier. Using the procedure described in previous exercise, measure the blood pressure of your partner under the following conditions, (a) while sitting (b) reclining position (c) standing position (d) a few minutes after physical exercise (e) after an untired 10-15 minutes physical exercise. Every time also note down the pulse rate.

Make a record of readings in the table form and by comparison draw inferences/conclusion.

Exercise No. 11

To demonstrate the presence of iron in the respiratory pigment, *i.e.* haemoglobin of blood.

Theory: Red colour of the blood is due to the presence of haemoglobin. Each haemoglobin molecule contains four haeme groups each with an iron atom at the centre. These iron atoms loosely combine with a molecule each of oxygen (O_2).

Procedure

(1) Take a sample of blood in a test tube and add 4% trichloroacetic acid (TCA) or 90% alcohol to it. A brownish red precipitate mainly of protein and a colourless solution will be left indicating association of pigment with protein and a colourless solution will be left indicating association of pigment with protein.

(a) Take a diluted sample of blood (with distilled water) in a test tube and add glacial acetic acid to it. Shake the mixture with a little ether and allow it to stand for sometimes. At this stage, three distinct layers or columns of liquid will separate out in the test tube. The upper layer will contain a reddish solution of ether in the pigment. Below this shall be noticeable a narrow zone of precipitated protein, which is devoid of colour (isolated from the protein by the action of glacial acetic acid or ether) and at the bottom a mixture of acetic acid and residual water.

(b) Evaporate a few drops of pigment solution to dryness. Ignite and extract with dilute HCl. Add a few drops of dilute potassium ferro-cyanide solution to this. The development of the distinct blue colour will indicate the presence of iron.

(2) Take 2ml of blood of frog or rabbit and evaporate to dryness. Add a small amount of fusion mixture (equal parts of anhydrous sodium and potassium carbonate) and heat until all the organic material has oxidized. Allow to cool then dissolve in a few drops of 10% ammonium thiocyanate. The development of a red colour due to the presence of ferric thiocyanate will indicate the presence of iron.

Exercise No. 12

Detection of the respiratory pigment called haemocyanin.

Theory: Haemocyanin is found dissolved in the plasma of a number of crabs, scorpions, gastropods and cephalopods. It is blue in the oxygenated state and colourless in the dissociated state. This pigment consists of a protein and prosthetic group containing copper.

Procedure

Take a few ml of the blood sample in a crucible and evaporate to dryness. Dissolve the dried material in a few drops of concentrated HCl. Add to this 2-3 large knife point of solid sodium acetate and then 1 drop of pyridine and 1 drop of 10% potassium thiocyanate solution. Mix well and decant the supernatant fluid in a clean test tube.

The development of a yellowish green colour will indicate the presence of copper in the blood, due to the reaction, which takes place in the presence of pyridine between copper salts and thiocyanate.

Add 5 drops of chloroform and shake well. The copper complex passes completely into the chloroform layer.

Exercise No. 13

To detect the presence of blood.

Material and reagents: Test tubes, microslides and coverslips, cotton, alcohol 90%, benzidine reagent, hydrogen peroxide, glacial acetic acid, spirit lamp or gas burner, microscope, *etc.* Benzidine reagent is prepared by dissolving, enough quantity of benzidine in glacial acetic acid in a test tube or a small conical flask to make its saturated solution. The preparation is facilitated when dissolution is brought about in a water bath set at 50°C.

Theory: It often becomes necessary to test for blood in the biological materials. The presence of blood in the urine, stomach contents or faeces is of great diagnostic significance. In medico-legal work, it is often important to determine not only the presence of blood but also the kind. There are two most commonly used tests to detect the presence of blood. These are as under:

(1) Guaiac or Benzidine test

It is very easy to perform this test. The sample consisting blood or a drop of blood is diluted in 2-3ml of distilled water in a small test tube. To this add about 2ml of benzidine reagent and gently shake the test tube to bring about thorough mixing. Now add 1ml of 3% solution of hydrogen peroxide. This results in the development of blue or green colour, which is characteristic of haemoglobin; thus ascertaining the presence of blood.

Note: This test is sometimes objected to in medico-legal work for the reason that other substances, such as milk, saliva and plant juices also give this test and so also does the enzyme catalase.

(2) Haemin Crystal Test

Theory: The blood pigment, called haemoglobin, is usually found in the blood. It is conjugated protein comprising a non-protein part called the haeme or haematin and globin. The haematin is composed of porphyrin group combined with iron. The chloride of haematin is known as haemin.

Procedure

Although the principle is the same but the test can be performed by using the differently modified procedure as under:

(1) Place one or two drop of blood at one end of the slide. To this add one drop of physiological sodium chloride solution (0.75%). Carefully evaporate to dryness over the flame of a burner. Add a few drops of glacial acetic acid and then cover it with a coverslip. Thereafter, heat it over a low flame of burner till effervescence or air bubbles come out. Remove it and quickly run from the sides of coverslip a few more drops of glacial acetic acid. Allow it to cool and examine your preparation. The formation of dark brown coloured rhomboidal or parallelogram crystals will not only indicate but will confirm the presence of blood.

(2) Take a drop or two of the uncoagulated blood at one end of the slide and make a smear of film of the blood with the help of another slide. Allow it to dry in the air. Thereafter scrape the blood film with the help of razor blade and collect the scraped blood in the form of powder at one end of the slide. To this add 1-2 drops of glacial acetic acid and a drop of normal salt solution. Put a cover glass over it. Now, hold the other end of the slide in between the index finger and thumb and heat it over the low flame of a spirit lamp or burner till the appearance of effervescence. Remove the slide immediately away from the flame and quickly add a few drops of glacial acetic acid from the sides of the coverslip without displacing it. Allow it to cool down to room temperature. Examine your preparation for the formation of the dark brown coloured rhomboidal shaped crystals. If such crystals are formed in your preparation then the presence of blood or blood in the sample is confirmed.

(3) This method should be followed when the sample of blood is available in the form of the old dried blood. In such a case, first of all prepare the powder of the sample if it is in the form of a dried lump. Place a little bit of glacial acetic acid and a crystal of NaCl or a drop of normal salt solution. Put a cover glass and heat it over the flame till the appearance of effervescence. Immediately remove it away from flame and add 2-3 drops of glacial acetic acid. Allow it to cool down and examine your preparation for the characteristic haemin crystals.

(4) This is the most simple and quick method. Take a drop of blood, draw a film and dry it. Scrape the film, collect the pow-

dered blood. Add a few drops of acetic acid and put a coverglass. Heat it gently, then cool and examine your preparations.

The rhomboidal shaped dark brown crystals as shown below will be visible.

Exercise No. 14

Detection of human blood from the blood of other animals.

Materials and reagents: Physiological salt solution, specific antiserum for human blood. This human antiserum is prepared by injecting the repeated doses of human blood serum into the blood of living rabbit. In due course of time the rabbit blood serum develops the power of forming a precipitate with a very dilute solution of human blood only and with no other blood of any animal.

Procedure

The test is also known as precipitation and compliment fixation test. Take the human blood and dilute it in the ratio of 1: 1000 with distilled water. Place the diluted sample in a test tube and add a few drops of human antiserum. Place it in an incubator for sometimes. The formation or appearance of a precipitate will confirm the presence of only the human blood.

Note: Readymade samples of this type of human antisera are available with chemists or medical stores. This technique is also used in forensic laboratories to ascertain the sample of blood of human beings.

Exercise No. 15

Determination of the haemoglobin percentage (Hb%) in the blood.

Requirements: Haemometer, N–HCl solution, distilled water, *etc.*

Theory: The red colour of blood is due to the presence of haemoglobin, an oxygen carrying pigment. It is usually found packed in RBC's but it is also found dissolved in plasma as in the blood of earthworm, haemolymph of the *Chironomus* larva. It is also found packed in other tissues, *e.g.* lining of the brood pouch of *Gastrothylax,* a helminth trematode endoparasite found in the rumen of cattle.

The determination of the percentage of haemoglobin is based on the conversion of the haemoglobin into acid haematin. The haematin thus produced is compared with the standard haematin by the apparatus known as haemometer.

This apparatus is referred to as Grower's Haldane haemoglobinometer or simply Grower's Haldane haemometer Figure 8.13. It has two vertical tubes, which are used as standard tubes for comparing the colour of human blood haemoglobin (or of any vertebrate like that of frog) solution (acid haematin) placed in haemoglobin tube or experimental tube. The haemoglobin tube is graduated in such a fashion that the graduated marks on one side give information in gm percentage while on the other side information about the percentage alone. A micropipette called Pasteur's sucking pipette is marked by 200mm (Figure 8.14).

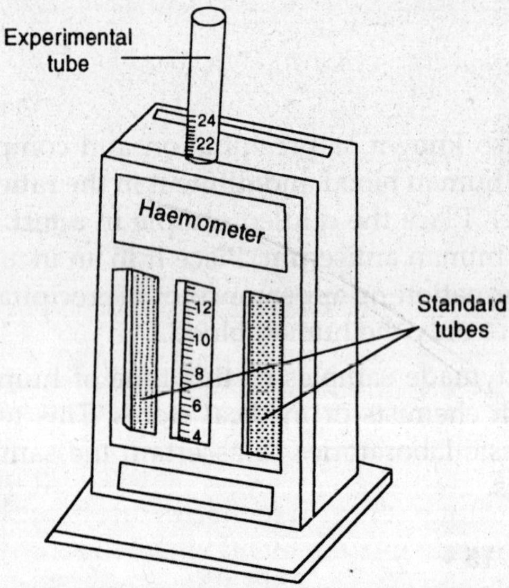

Fig. 8.13. Grower's Haldene Haemometer

Procedure

Proceed step by step to ensure satisfactory results.

(1) First of all make sure that both the haemometer and the Pasteur's sucking pipette are clean and dry.

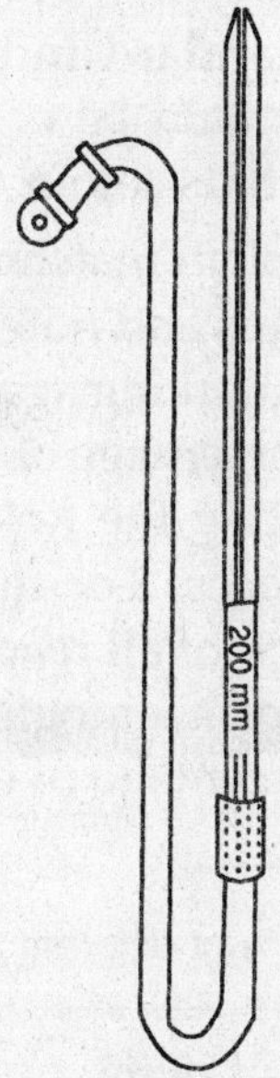

Fig. 8.14. Pasteur sucking pipette

(2) Fill up the haemoglobin tube with 20ml of N-HCl upto 20th mark on percentage side.

(3) Either prick the ring finger of your left hand with a well-sterilized needle for easy oozing of blood (in case of human blood) or prick the heart of frog with a needle (in case of frog's blood). Suck up the blood in micropipette exactly upto the mark of 20 in the middle of pipette. The excess of blood beyond the tip or outside the tip should be wiped off with a filter paper, taking care that volume of blood in pipette does not fall below mark 20.

(4) The blood from pipette is transferred or discharged into the tube containing N-HCl. To ensure complete transfer of blood from pipette, N-HCl is again sucked up in the pipette upto mark 20 and returned to the mixture in the haemometer tube. Thoroughly mix the blood with N-HCl in haemometer tube by constant stirring with a glass rod. This process of stirring should be continued for at least 10 minutes. During this period, the conversion of haemoglobin into haematin will take place and a distinct dark brown colour will be developed. This will ensure the complete formation of acid haematin.

(5) After this, start diluting the mixture, gradually adding drop by drop either water or N-HCl and go on matching the colour

of the diluted medium with that of the stand tubes of haemometer working as standard indicators. The simultaneous stirring of mixture should be continued while diluting the mixture so as to match its colour with that of the stand tube.

(6) When satisfied with colour matching, take out the haemometer tube from the apparatus and note down the percentage reading on the graduated marking against the top level of the mixture. This gives the percentage of haemoglobin present in the blood. Before presenting the final result regarding the percentage, repeat the process 3 or 4times and take reading every time. Then the average of all readings should be reported as the final percentage of haemoglobin present in blood. This will minimize the percentage of error, which might have occurred in one reading.

Table 8.5: Haemoglobin % at different ages of male and female.

Age	Male % in gm	Female % in gm	Remarks
Birth to 1 ½ months	15.5	15.5	
3months to 1yr	12.5	12.5	Haemoglobin percentage may be reduced due to mortality of RBC & poor contents of iron & milk
From 2-15yr	14	14	
Adults	15-17	12-15	Low in female because of menstruation
Old age	14.7	14.7	Because of menopause in female

Precautions

(1) Suck up the blood in Pasteur pipette exactly upto the marking 20.

(2) Avoid entry of air bubble in the pipette, if so happens discard the blood and again suck up. This could be avoided if the pipette is clean and dry and at the same time its nozzle should be as much inside the drop as possible.

(3) Excess of blood adhering to the tip of the nozzle should be wiped off with filter paper.

(4) The colour matching should be done in a perfect and accurate manner.

Since the haemoglobin percentage is dependent on a number of factors, such as, age of the person, sex, climate, *etc.* (Table 8.5). So, a tentative scheme is outlined in the table given below; otherwise the values roughly vary between 14gm/100ml to 18 gm/100ml.

Exercise No. 16

Determination of the oxygen carrying capacity of blood.

In this exercise the percentage of haemoglobin is required, so the procedure remains the same as in earlier exercise. From the haemoglobin percentage determined, the oxygen carrying capacity can be calculated as under:

Since it is well mentioned in the literature that 1gm/100cm^3 of haemoglobin (the respiratory pigment) reversibly combines with 1.34cm^3 of oxygen. Therefore, the 14gm/ 100cm^3 of blood will combine with14 x1.34cm^3 of oxygen. From this it can be concluded that with an increase in the percentage value of haemoglobin/cm^3 of blood, the oxygen carrying capacity of blood is increased.

Exercise No. 17

To find out the absorption spectrum of a oxygenated and deoxygenated (reduced) haemoglobin.

Requirements: Pupillary hand spectroscope, test tube, blood, Stroke's reagent.

Theory: If white light is passed through a solution of a pigment, certain wavelengths are absorbed more strongly than others. If the white light that has been modified and passed through the prism of the spectroscope, the familiar spectrum of white light will be found to be blocked out in certain regions. The position and density of these absorption bands running transversely across the spectra are characteristics for the type of the pigment. The changes in the chemical state of the pigment, which produces a change in colour, reflect in the position of the absorption band by using a pupillary hand spectroscope (Figure 8.15). In this way, it is possible to make sensitive comparisons between respiratory pigment of different animals or of the same respiratory pigment under different conditions of loading and unloading with O_2. The units used in the spectroscope are m/μ & A° units. Instead of m/μ, the unit nanometer 'nm' is in current use.

The absorption spectra, which are obtained when white light is passed through Hb solution or closely related derivatives, are useful in distinguishing them from one another: HbO_4 or diluted atrial blood shows three absorption bands.

(1) A narrow band of light absorption at a wavelength λ= 578nm.

(2) A wide band of wavelength λ= 542nm.

(3) A band of wavelength λ = 415nm.

CO_2 combines with Hb even more rapidly than does O_2. The carboxy haemoglobin (CO Hb) shows two absorption bands, the middle of the first at λ = 517nm and the second at λ = 542nm. Combination of Hb with HgS or HCN also gives characteristic absorption spectra. This is valuable method of detecting these compounds in the blood of individuals poisoned with HgS or HCN.

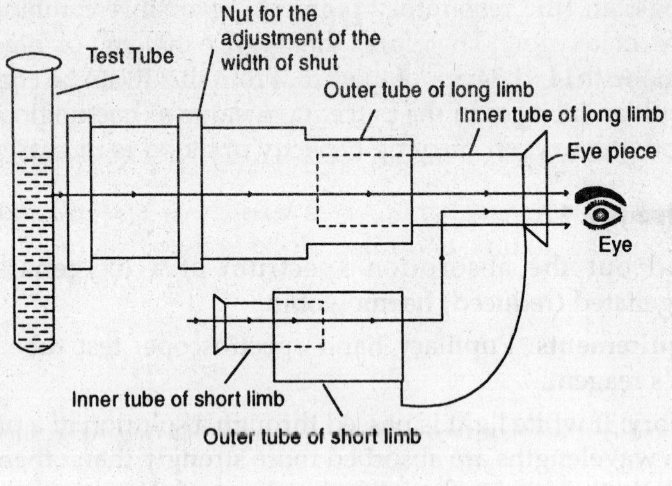

Fig. 8.15. Pupillary hand spectroscope

Setting up of Pupillary hand spectroscope

This apparatus is used in studying the absorption of various pigments present in the blood and very simple for the examination of blood spectra.

This instrument consists of two limbs (a) with a short and narrow limb, (b) a long and broad limb. Each limb consists of two telescopically sliding tubes. In broader tubes, there is a nut for the adjustment of the slit and in the narrow tube is fitted a scale. There is an eye piece through which one can see the spectrum and the scale simultaneously.

Adjustments of Pupillary hand spectroscope

(1) Close the slit point of the instrument towards the source of the light and open the slit very slightly. Now, focus the Fraunhauffer's lines of the solar spectrum. These Fraunhauffer's lines appear in fine vertical block lines across the spectrum. All these lines are noticeable only when a very narrow slit is used.

(2) Focus the wavelength.

(3) Set the scale with reference to Fraunhauffer's line. It is much better to use the solar line than the lines obtained by means of a sodium flame.

Procedure

(1) Clamp the pupillary hand spectroscope on a hand and adjust it in such a way that the spectrum of ordinary or solar light coming through the objective line is clearly visible. Reading on the scale should be carefully recorded.

(2) Take the blood in a test tube and dilute it with distilled water.

(3) Now, take two separate tubes labelled as A and B and transfer equal volume of diluted blood in these test tubes.

(4) Take test tube B and further dilute the diluted blood in it with the addition of more distilled water or Stroke's reagent.

(5) Place the test tube A in front of the objective line of spectroscope and observe two bands of oxyhaemoglobin in the yellow green region. Note down the readings of the absorption spectrum of oxyhaemoglobin.

(6) Similarly, take test tube B in front of the objective line of spectroscope, observe the band and note down the readings and compare with Table 8.6 for normal blood.

Note:

(1) In reference to the microwave length, mμ, the presence of two absorption bands (525 mμ-555 & 565-585 mμ), in yellow-green regions indicate the presence of oxy-haemoglobin or oxygenated haemoglobin, whereas a single absorption band (540 mμ-585 mμ) indicate the presence of deoxygenated or reduced haemoglobin.

(2) Units used in spectroscopy are mμ (millimicron=0.001μ) and the Angstrom unit (=0.1 A°).

Table 8.6: Wave length of oxygenated and deoxygenated Hb.

State of Hb	Number of bands	Wavelength at which the band appeared
Oxygenated Hb	2 bands	1st 550nm 2nd 600nm
Deoxygenated Hb	1 band	600nm

(7) Similarly, take test tube B and notice that the colour changes to a violet red. Now, place the tube against the source of light entering the spectroscope and observe that two bands seen previously have now united into one broad-spectrum band. If the test tube is shaken for a little while in air or O_2 is passed into it, again the two bands of oxyhaemoglobin are noticeable. Note down the readings of the absorption spectra of reduced haemoglobin.

Preparation of Stroke's reagent

It is a solution containing equal volumes of 2% ferrous sulphate and 3% tartaric acid. At the time of use, a small amount is placed in a test tube and ammonia is added until a precipitate at first undergoes immediate dissolution. This produces ammonium ferrotartrate, which is a reducing agent.

Further details of the absorption spectra of blood under different physiological conditions are revealed by spectroscopic examination.

(a) Oxyhaemoglobin

The haemoglobin that has not been exposed to air is purple in colour but when it is exposed to air, it takes up molecular oxygen and becomes changed into oxyhaemoglobin, which is scarlet in colour. This change is readily noticeable in meat market. The meat of freshly slaughtered animal is purple due to the presence of reduced form of haemoglobin inside the tissue. Shortly afterwards it takes on the scarlet colour due to oxidation by air.

When examined with hand spectroscope, two bands of oxyhaemoglobin in the yellow-green regions are clearly noticeable.

(b) Reduced Haemoglobin

Add small amount of Stroke's reagent to a sample of blood in a test tube. The colour changes to a violet red. Observe that two bands seen in oxyhaemoglobin unite into a broad band. If the same mixture is shaken in the air or O_2 is passed into it, the two bands of oxyhaemoglobin again are clearly perceptible.

(c) Carboxyhaemoglobin

If oxyhaemoglobin is treated with carbon monoxide, the carbon monoxide will replace oxygen forming carboxyhaemoglobin, which is bluish red in colour and is much more stable than oxyhaemoglobin. When the blood becomes 60% saturated with CO, death of person occurs. Some of the symptoms of CO poisoning are headache, dizziness, dim vision, nausea, unconsciousness and convulsions.

For spectroscopic examination, take in a test tube some blood and pass the gas of a burner with a nozzle. Observe the colour changes to a carmine tinge. The two bands like that of oxyhaemoglobin appear but are nearer to the violet region of spectrum. Add a little of Stroke's reagent and notice that no alteration of bands take place.

(d) Methaemoglobin (Neutral)

When oxyhaemoglobin is treated with potassium ferrocyanide, it loses oxygen and its iron is oxidized to the ferric state, forming methaemoglobin, which is brown in colour. The brown colour of the meat under certain conditions may be accounted for by the presence of methaemoglobin.

Add a few drops of freshly prepared 10% solution of potassium ferrocyanide to the blood sample in a test tube. Observe the development of brownish tinge. On looking through the spectroscope, notice one band in red region and two faint bands of oxyhaemoglobin. Now, add a drop of Stroke's reagent and observe that oxyhaemoglobin absorption spectrum is followed by that of haemoglobin.

(e) Methaemoglobin (Alkaline)

Add a few drops of freshly prepared 10% solution of potassium ferrocyanide to the blood sample in a test tube. To this add a few drops of ammonia and notice the development of a reddish colour. On looking through the spectroscope observe two bands in different position than oxyhaemoglobin and one faint band nearer the red region.

Tabulate the range of the wavelength of each band of the spectra in each case.

9

Kymography

THE KYMOGRAPH (THE CLASSICAL TOOL OF PHYSIOLOGISTS)

Kymograph is a mechanical recording instrument (Figure 9.1), which no doubt in this age of mechano-electric transducers and electronic recording devices seems to be 'old fashioned', but to restore the lost confidence in this classical tool of physiologists and to apprise the beginner to realize the tremendous potential of this instrument as a universal recording device, the under mentioned salient features are noteworthy.

Basically, the instrument is used to record mechanically the muscular activity, *i.e.* contraction and relaxation phases. Each living cell is excitable. It can be initiated by some external agent called the stimulus. The stimuli may be of the following types: (a) Thermal (b) Chemical (c) Electrical, and (d) Mechanical.

Effect of these stimuli varies from cell to cell. Muscles show most pronounced effect. They respond to the stimuli by contraction followed by relaxation; both of which are mechanically recorded. By these tracings, the intensity, frequency and durations of contraction and relaxation phases of muscles can be studied. In all such types of tracings, the organ under investigation is connected to a lever by means of a thread, so that when it contracts; the lever is moved from its resting position and the stylus attached at the free end of the lever inscribes the tracings on the smoked kymograph paper, mounted on a rotating drum.

The most important feature of kymograph nowadays most commonly used in physiology laboratory is a metal cylinder mounted

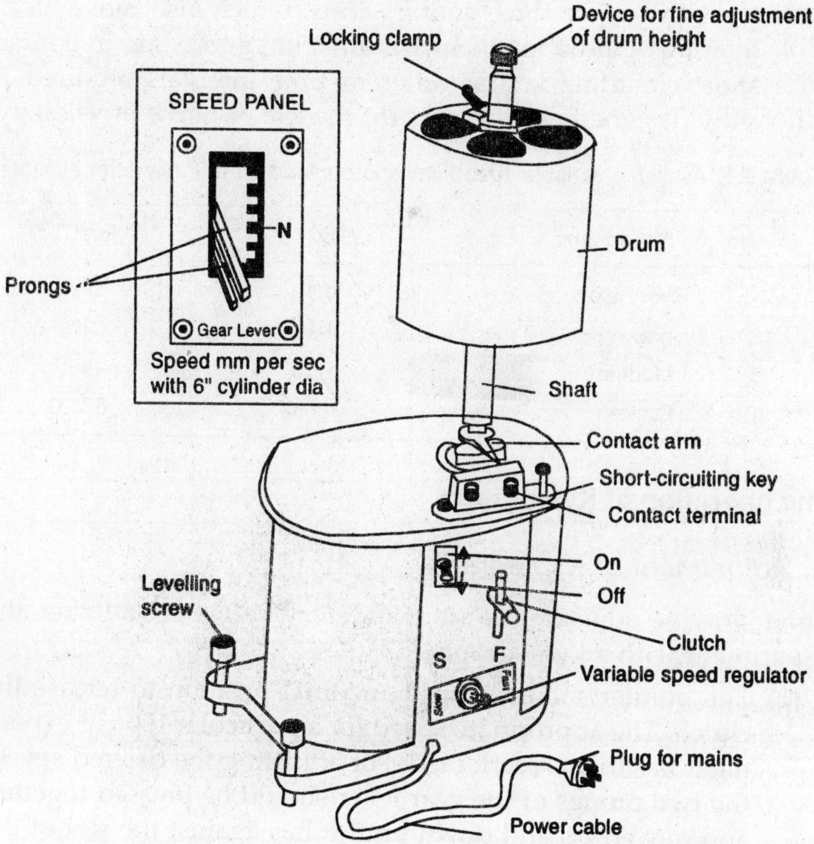

Fig. 9.1. Kymograph apparatus with speed panel shown separately

on a rotatable spindle or shaft that itself is rotated by an electrically operated motor. The motor is enclosed in a gearbox that itself is mounted on leveling screws. The speed of rotation is regulated through gear arrangement provided on one side of gearbox. Usually four to six ranges of speeds depending upon the number of revolutions of the cylinder are provided. In addition, a variable regulator permits two variation marked against each slot. This is made possible by shifting the variable regulator from the slow (S) position to fast (F) position. This variation in speed at fast as against the slow is approximately double as shown in table 9.1.

On the other side of the gearbox, there is one ***on/off*** switch and a gear lever. When the gear lever is in neutral position, the shaft remains stationary, *i.e.* it does not rotate. In addition to various arrangements on left and right side of the gearbox, the top base of gear box is provided on right side, with two contact terminals for short circuiting arrangement, while the base of rotating spindle or shaft with two or three contact arms, which also move along with rotating spindle in clockwise direction and come in contact with short circuiting key at different time intervals required in differently designed experiments for muscle or nerve physiology.

Table 9.1: Available variable speed limits in mm/sec on a 6" diameter cylinder.

S. No.	Speed limit	Slow 'S'	Fast 'F'
1.	Very slow	0.12	0.25
2.	Slow	1.25	2.5
3.	Medium	12.5	25.0
4.	Fast	320.0	620.0

The operation of Kymograph

A. Adjustment of Kymograph

The following adjustments are mandatory before conducting any experiment with a kymograph.

(1) The pointer should be in horizontal position to release the clutch. The appropriate speed for a particular type of experiment has to be ascertained. For adjusting the desired speed, the two prongs of the gear lever should be pressed together and moved up and down until it lies against the slot of the chosen speed. In this position the two prongs should be released. The variable regulator should also be adjusted accordingly for slow or fast speed requirement.

(2) With the help of a spirit level, using the leveling screws, adjust the gearbox in a perfectly horizontal position.

(3) Connect the power cable with main switch taking care that the on/off switch is in off position.

(4) Now, the switch should be put in on position and turn the clutch so as to remain in vertical position. The drum will start rotating clockwise at the adjusted speed. This ensures that the kymograph is ready for use.

(5) Stop rotating the drum by bringing the clutch in horizontal position. The rotation of the drum should not be stopped by using 'on/off' switch.

(6) In manual operation of the drum, the gear lever should be

affixed at neutral position on speed panel. Do not rotate the drum while motor is in gear.

B. The application of kymograph paper on drum

All types of paper are not suitable, for this purpose only glossy white paper (glazed paper) specially designed for this purpose is used. The drum is removed from the kymograph shaft by unlocking it. Its circumference is measured with the help of a thread. Thereafter, the paper is cut from the paper roll about 1 or 2 cm more in circumference so that one end overlaps the other when the cut out piece is rolled around the drum. With a view to apply the cut paper strip on to the drum, lay it on the working table with its glossy surface down in contact with table. Apply glue paste on the left hand side 1 to 1.5cm from the cut margin of the strip. While applying glue paste, it should be ensured that on rolling the paper on the drum, the glued surface should be fixed on the surface of paper on other end and should not come in contact with the drum surface. Lift the two ends of the paper strip up and join them on top of the drum. Orient strip and drum in such a way that the top of the drum faces away from you. The ends of the paper strips must then overlap so that the end of the left half lies over that of right half. Press the ends of the left half of the paper strip over it. The actual procedure is displayed in Figure 9.2. As the paper dries, it stretches and fits on the drum surface tightly so that the paper cylinder does not slip off automatically. The joint is arranged in such a fashion that the writing stylus be not caught by the edge of the paper joint, when the drum rotates in the normal clockwise direction. Further onwards the drum should be handled very carefully avoiding any denting or distortion of the surface of paper.

C. Smoking the kymograph paper on drum

This involves the application of a thin uniform film of soot on the surface of glazed paper already affixed in the drum. Hold the drum on a horizontal shaft of the smoking stand and place it on the stand. Light the wide wick of the kerosene oil burner. This will provide a wide range of flame. Place the burner below the drum and start revolving the drum quickly and briskly so that a uniform speed of revolutions is maintained throughout the smoking. Also keep on shifting the flame back and forth the paper. This will help to make a thin uniform layer of soot on the paper. While smoking, care should be taken not to place the flame very close to

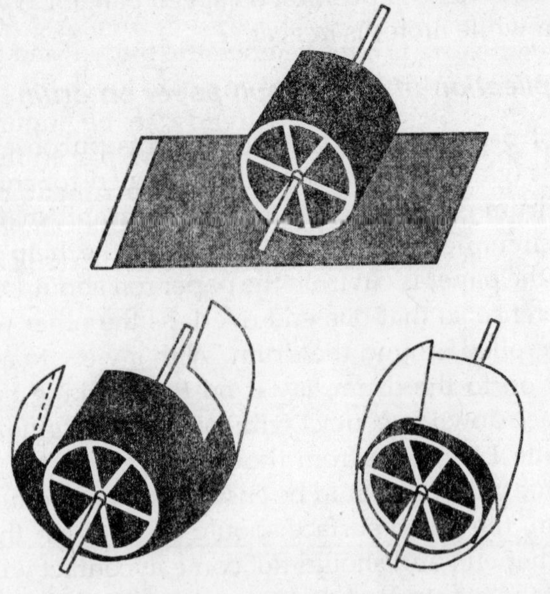

Fig. 9.2. Diagram illustrating the procedures involved in the application of kymograph paper on drum.

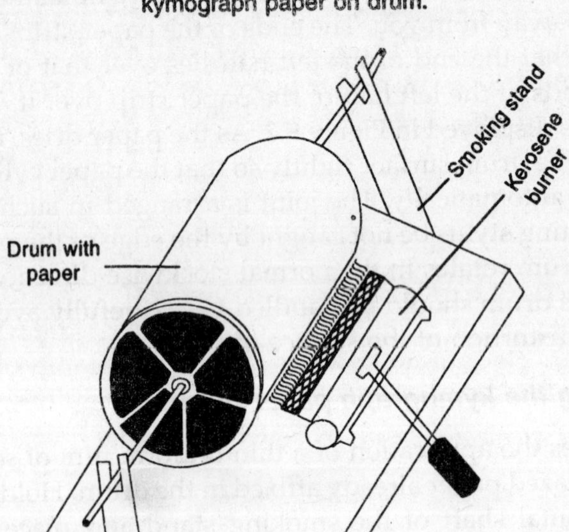

Fig. 9.3. Diagram illustrating the procedure involved in smoking kymograph paper.

the paper else it may be burnt or the paper will stick. Besides giving trouble when record is removed later, it damages the drum. As soon as the smoking is over, remove the burner and stop rotating the drum. Every care has to be taken not to touch the paper surface afterwards. The drum is now ready to be mounted on the kymograph shaft; at this stage it should be placed in a safe place before being mounted on shaft for any experiment; because most often by sheer negligence or accidentally the smoked paper is smudged. The complete arrangement of smoking is illustrated in Figure 9.3.

D. The types of recording levers

These are of different types and are named according to their use and function as for instance the heart lever, muscle lever. Irrespective of the type of lever, each is provided with an arm carrying at its distal end a writing stylus.

E. Adjusting the lever

Before performing any experiment with the freshly smoked drum, it is very important that the lever and writing stylus is placed at the proper position. For this a roughly smoked drum should be mounted on the shaft of the drum and the lever be adjusted so that it moves in the plane of the tangent of that point at which the writing stylus just touches the circumference of the drum as depicted in Figure 9.4, *i.e.* the lever be set at right angles to the radius of the drum at the writing point. It should also be ensured that the lever always remains directed to the left, simply because the drum revolves in a clockwise direction. Apart from this, it should be ensured that the shafts of kymograph holding the drum and kymograph stand holding the lever are absolutely vertical. Even so the tip of the stylus should never leave the paper as it moves up and down from the original horizontal position in which it touches the paper.

F. Final recording on smoked paper

When all adjustments have been made satisfying all the terms and conditions, put the freshly smoked drum on the shaft of the drum. Adjust its position and lock it at this place by pressing the locking clamp towards the shaft. This will not allow the slipping of drum vertically downwards during recording. Bring the stylus tip near the smoked paper so that it just touches it. If required make a base line by rotating the drum manually in clockwise direction. All

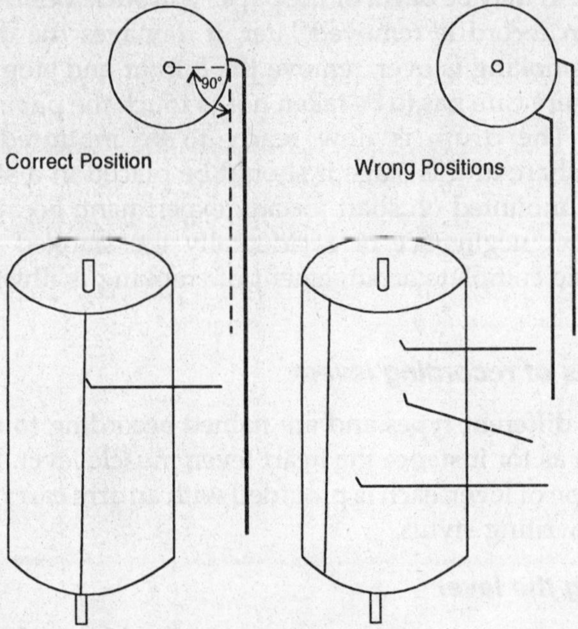

Fig. 9.4. Diagrams illustrating the correct placement of the writing stylus and some common mistakes in setting up and adjustment of the writing stylus.

recordings should always be started from the right side of the joint. Start rotating the drum by putting the clutch in vertical position maintaining a sharp and constant vigil over the recordings so that the repetition and overlapping may not take place. When the recording of tracings is over, stop the moving drum by bringing the clutch in horizontal position. During recording, the motion of the writing stylus over the smoked surface of kymograph paper is inscribed in the form of a continuous white line and at the same time, the changes in position of the writing stylus are recorded in relation to the base line already drawn. The horizontal axis or base line corresponds to a time axis, while a chosen distance between two points on a base line corresponds to a definite time interval, which is proportional to the speed of the rotation of the drum. For this, before removing the drum after recording make a time scale with the help of an electrically operated time marker. The time scale recording should be either below or above the tracings of the experiment. The time scale is recorded on the same speed at which the experimental tracings are recorded. When everything has been completed, remove the drum from the shaft by unlocking it and put it on the table so that its topside, that is upper side, carrying locking device faces upwards towards you.

G. Removing of smoked paper

Tightly hold the paper at the upper end with the help of two fingers of left hand around the joint and make a vertical cut along the joint by means of a sharp razor or scissors and lay the paper horizontally on smooth, neat and clean wooden table in such a way that the surface bearing the record of tracings faces you. Pin up the four corners of the paper strip. While removing the paper, take care that the surface of paper does not come in contact with anything or else the tracings will be spoiled/tampered/mutilated. The procedure is illustrated in Figure 9.5.

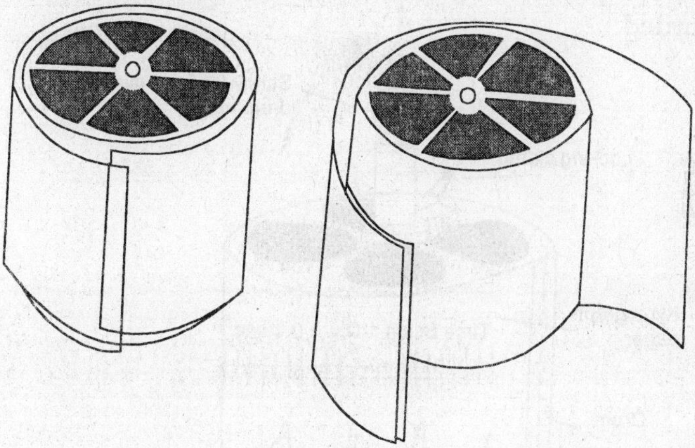

Fig. 9.5. Diagram illustrating the procedure of removing paper.

H. Labelling of the tracings

This is done with the help of a penholder carrying a dry nib without ink or a pointed blunt needle. Label at appropriate place the following: the name of experiment, speed of rotation, the value of each division of time scale, your name, date of recording, *etc.* Besides each part of the tracing should be specifically labelled in accordance with the design of experiment and nature of results required. This will be made clear in subsequent specific exercises. An illustration is presented in Figure 9.6.

I. Fixing the tracings for permanent record

In this process, as illustrated in Figure 9.7, the sooted paper strip bearing the labelled inscribed traces of recorded lever movements is fixed in a fixing solution to make a permanent record. For this purpose, the paper strip is submerged in a fixative contained in a tray. Hold the two ends of the strip with the tracing bearing side upwards. Gently give a curvature and dip the central curved position of the strip in the fixative so that some fixative flows over it. Gradually move the strip, keeping it submerged from first one way towards the left, then other way towards the right. Repeat the process twice. Thereafter, hold only one end of the strip, leaving other end pointing towards the tray so that excess of fixative flows down in the tray. Hold the strip on a thread string with the help of paper clips or gem clips and leave it for air-drying (Figure 9.8). Transfer the fixative in bottle again. When it is completely dried, it could be used as a permanent record in the practical notebook. Only the best portion of the record should be maintained and rest be discarded.

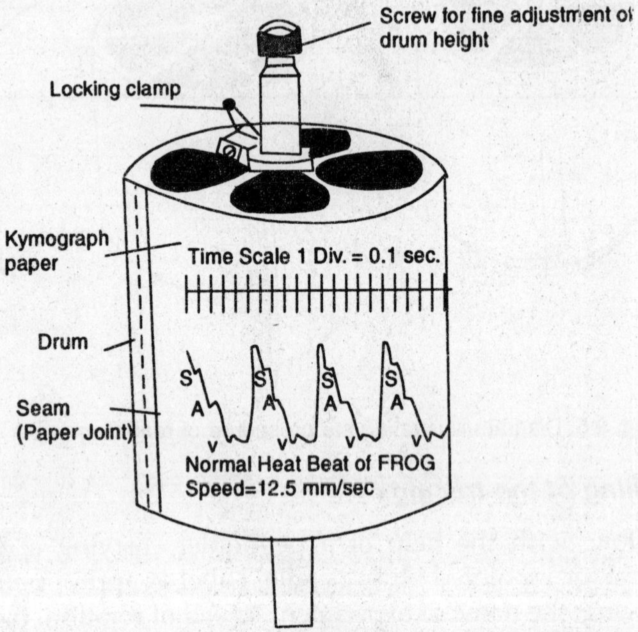

Fig. 9.6. Diagram illustrating the recording of the tracings of the normal heart beat of frog by Kymograph.

The most widely and commonly used fixative consists of a solution of shellac in rectified spirit. It is easily prepared by dissolving 250g of granular shellac powder in 500ml of rectified spirit (96% denatured) and add 2.5 to 3.0ml of castor oil. The addition of castor oil makes the dried film of lacquer flexible and prevents the development of cracks when the paper is rolled up for storage for longer durations. Apart from this addition of oil imparts a shining glaze to the whole record.

Fig. 9.7. Diagrams to illustrate the procedure involved in fixing the Kymograph paper.

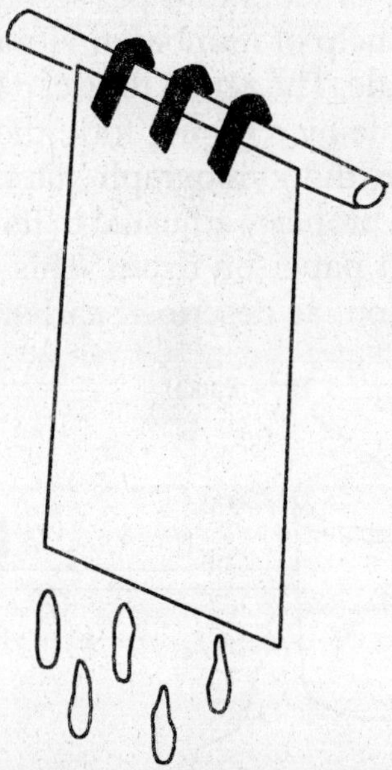

Fig. 9.8. Diagram to illustrate the procedure involved in drying the Kymograph tracings for permanent record.

Other Accessories

1. The Time Marker

These are of different ranges but all are electrically operated. For example, direct writing time marker, with ranges varying from $^1/_5$ sec, 5 sec and 30 sec, 1 sec and 10 sec. These time markers are used to make time scale on drum. As the movement of paper is based on the intensity of stimulus, frequency or the period of some physiological activity, the time scale marking becomes altogether an important feature of any experiment. The time marker (Figure 9.9) is firmly held on a kymograph stand and is connected with main. Its writing stylus is brought near the smoked paper and set in proper position. It is started and the drum is allowed to rotate at the same speed at which the movements of the lever associated with any activity were recorded. In the absence of a time marker being operated by the Galvanic current (constant current) as above, the signal magnet type being operated by the Faradic current can be used. Nowadays this type is not in use in most of the laboratories because of the availability of the AC main operated time markers. A time scale can also be made using tuning fork (Figure 9.9b), which is provided with a writing stylus at the end of one of its prongs. The frequency of vibrations per second is marked on such tuning forks with the help of number of vibrations covered in any activity on the time scale. The actual time can be readily calculated. For making a time scale by a tuning fork, the posterior rod of the fork is firmly held on the kymograph stand with the help of a clamp. Thereafter, it is properly adjusted in its proper position with respect to the smoked paper on drum. This is made close to the joint in tangent position, as described earlier. Using a starter, the

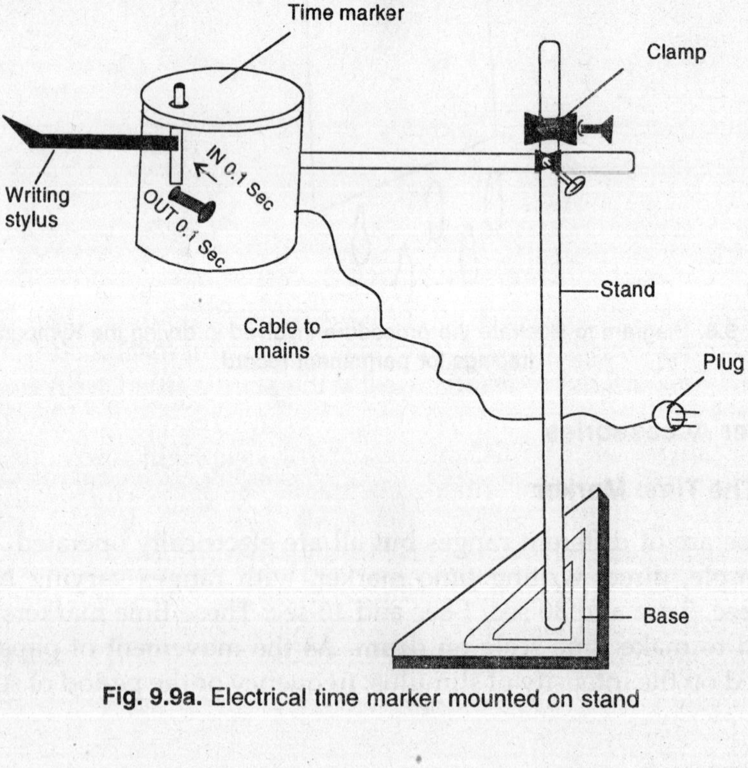

Fig. 9.9a. Electrical time marker mounted on stand

two prongs are pressed together. The fork starts vibrating. The starter is immediately removed and the drum is set in motion at the same speed at which the activity or movements of the lever of any activity were recorded. The most commonly used tuning fork bears the marking of 100 vibrations per second.

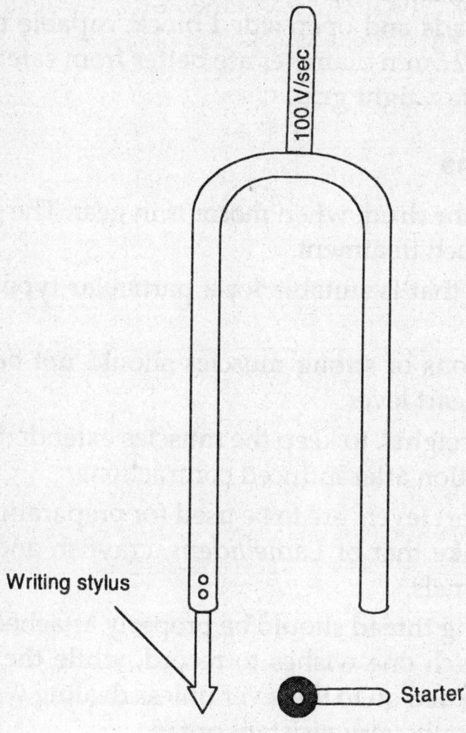

Fig. 9.9b. Tuning fork.

2. Kymograph stands

These are also of different types but the main feature of all is that they have a sturdy, stable machined bases and machined upright, shaft for perfect verticality. Mostly the stands fitted with fine adjustment devices are in common use. Some commonly used stands are plain stand, 38cm, 45cm high, stainless upright, and adjustable stand with quick tangential adjustment. The stand with 38cm high stainless upright, and large adjustable screw stands are, which is most popular and consists of a heavy lathe slide carriers, a dovetailed machined platform for fine movements. The to and fro movement of the vertical upright is brought about by a screw head, especially for fine adjustments.

3. Clamps and Boss Heads

These are also of different types and sizes but the main feature of each is that the writing levers, time marker, including tuning fork, the muscle chamber, *etc.* can be clamped to the stand with any type. Usually these are either of ordinary type like the open right angle clamp or boss head type. In comparison to ordinary type clamps, the boss heads and open sided block, capable of holding rods from 3mm to 12.5mm diameter are better from safety point of view as they provide a tight grip.

General precautions

(1) Never rotate the drum when motor is in gear. The gear tends to jam after such treatment.

(2) Select a lever that is suitable for a particular type of experiment.

(3) The contractions of strong muscles should not be recorded with a light heart lever.

(4) Use counterweights, to keep the muscles extended and to facilitate relaxation after induced contractions.

(5) Only light heart levers are to be used for preparations of delicate hearts, like that of *Lamellidens*, crayfish and other organs and animals.

(6) The connecting thread should be properly attached to the organ from which one wishes to record, while the other end should not be tied on to the lever unless dealing with a tough and mechanically very resistant organ.

(7) The drum should be set at the lowest possible position on the shaft so that it can later be raised when first round of recording is over.

(8) Before beginning the recording, make sure that the drum is securely fastened, *i.e.* properly locked and is not in the risk of slipping.

(9) Adjust the writing lever so that its tip just touches the paper near the upper margin of the drum on the right of the joint where the two ends of the paper strip have been glued together.

(10) It should be ensured that the arm of lever is in a horizontal position when the preparation is attached to it in its normal and resting position and the thread connecting the organ with

the lever must be vertical, otherwise great distortion is introduced into the record and it is valueless.

(11) It should always be borne in mind that the recording is curvilinear, that is the y-axis is represented by a semicircle, the radius of which is given by the length of the writing arm of the recording lever.

(12) The plane of up and down movements of the writing lever should be parallel to the tangent through the drum radius that goes through the lever.

(13) While making adjustments, the thread slack that connect the loaded arm of lever to the preparation should be left away, else unknowingly the jerky movements might tear and injure the preparation.

(14) Adjust the speed of the motor so that the record is of a reasonable length. Slow phenomenon obviously requires much slower drum speed than fast phenomena. No record should cross the joint in the paper. Before starting the experiment, always calibrate the speed of the drum. Although the time markers available by and large give quite accurate results but there is no harm in marking off suitable time intervals, such as 1min, 0.5min, 10sec or even 2sec in case of fast turning drums.

(15) If the movements of the lever recording the physiological activity is downwards as in case of experiments on heart beat of frog, then the time scale should be drawn above the tracings and if the movement of lever is upwards as in case of experiments with nerve-muscle preparations, the time scale should be drawn below and that too below the base line.

(16) Remove the paper by slitting it down with a scalpel or pointed needle of forceps, cut outwards, never into drum. Before fixing the tracings for permanent record, the tracings should be labelled very accurately and neatly. In addition to labelling of tracings, the record should also bear the name of the experiment, the name of the student and the date. All labelling should be done only when the paper is removed from the drum and placed horizontally on the smooth top of the experimental table.

(17) Only when all the precautions and conditions are rigidly and consistently followed, then it becomes quite easy to interpret each activity most accurately.

(18) The kymograph, like the camera, is a recording instrument. If for some reason, a record is unsatisfactory, repeat the operations and get a better record, which shows satisfactorily the phenomenon under investigation.

(19) The final record must be recorded continuously and not pieced together from separate records.

10

Physiology of Heart

INTRODUCTION

The heart of frog is composed of the muscles called the cardiac muscles. These are medium fast muscles and are responsible for the autorhythmicity of the heart. The cardiac muscles are spread over the entire section of heart, *i.e.* the sinus venosus, auricle and ventricle. These muscles contract simultaneously with considerable speed. The contraction is followed by the relaxation. The phases of contraction and relaxation of muscles of each chamber are well coordinated and regulated, which may differ under varying conditions of functional stress and physico-chemical environmental strains. The contraction and relaxation of cardiac muscles and effect of drugs and variables may be readily recorded by electro-mechanical device called the kymograph. This chapter deals with the exercises on the activity of heartbeat of frog under normal and varying physiological conditions to study various phenomena of frog's heart activity.

SOLUTIONS AND REAGENTS REQUIRED

Preparation of Ringer's solution for frog (According to Zoethort)

1ml of 1% solution of $CaCl_2$ (calcium chloride)

0.75ml of 1% solution of KCl (potassium chloride)

1ml of 1% $NaHCO_3$ (sodium bicarbonate)

97.25ml of 0.7% NaCl (sodium chloride)

Ringer's solution (According to McLung)

Distilled water	100ml
NaCl	0.8gm
$CaCl_2$	0.02gm
KCl	0.02gm
$NaHCO_3$	0.02gm
Dextrose (optional)	0.1gm

Exercise No. 1

To record the normal rate of the heartbeat of frog.

Apparatus: Kymograph apparatus, cardiac lever, heart clip, time marker, kymograph stand, chamber for holding the heart lever, dissecting instruments, dissecting tray, frog board, normal salt solution or Ringer's solution, urethane, hypodermic syringe, rough cloth, *etc.*

Theory: The heart muscle contracts rhythmically and shows an alternate contraction and relaxation. The contraction is termed ***systole*** and the relaxation as ***diastole***. Both the nerves and muscles constitute the conducting system; therefore the heart of frog is neither purely myogenic nor exclusively neurogenic in nature but is an intricately related union of the two, as against the earlier belief that the heart is purely myogenic and independent of the central nervous system.

The heart (Figure 10.1) consists of sinus venosus, two auricles, one ventricle and a bulbous or conus arteriosus. The heartbeat starts from the sino-auricular node (SA node), which is represented by a white crescentric line between the sinus venosus and auricles. The heart tissue or muscle has the property of innate rhythmicity or myogenicity, which is most highly developed in the sino-auricular node. S.A. node is capable of generating electrical impulses at regular intervals and controls the rate of heartbeat; hence it is called a ***pacemaker***. It does not require any stimulus from the nervous system. When the latter (*i.e.* auricles, ventricle and conus arteriosus) are cut off from the influence of sinus, they stop beating for a short while but soon afterwards they again start beating at a rate of their own. This is known as the ***idio-ventricular rhythm***. Normally the sinus venosus contracts first, it is followed by auricles, then ventricle and lastly, the conus arteriosus. The wave of contraction spreads so rapidly over each of the heart chambers, it appears that these contract simultaneously.

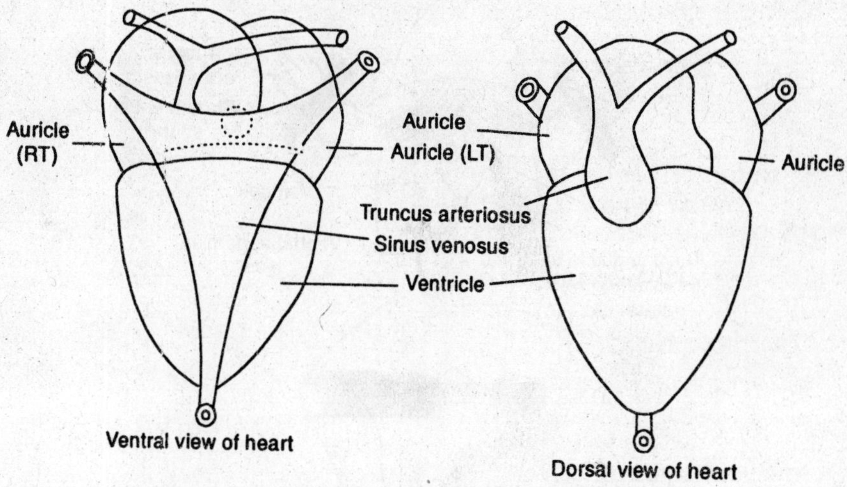

Fig. 10.1. Ventral and dorsal view of heart of frog.

Procedures

Strictly follow the details of each step mentioned below with a view to arrive at most satisfactory conclusion and accurate results.

(a) Preparation of the heart

Hold a frog very tightly with your hand using rough cloth else it might slip away. Narcotize the frog by injecting in dorsal lymph sinuses 2 or 3ml of 25% or 40% urethane solution in distilled water by means of a 5ml hypodermic syringe just underneath the skin. Narcotization may also be brought about by double pithing, *i.e.* by inserting a metallic pointed needle into the foramen magnum and rotating it either way. Place the frog in the dissecting tray and cut away the skin over the sternum and make an inverted-V-shaped window through the sternum. For this, two cuts into the sternum by means of a bone cutter are required, as shown in Figure 10.2, taking care that the underlying heart is not injured during bone cutting operation. Care should also be taken not to cut the ventral abdominal vein. The heart enclosed in the pericardium will be exposed. Carefully remove the pericardium, to expose the heart right from its base. The heart should not be grasped by means of the forceps.

(b) Mounting of frog on the frog board

Gently transfer the frog on a frog board and fix it by pulling the fore-limbs through the elastic bands placed at four corners of the

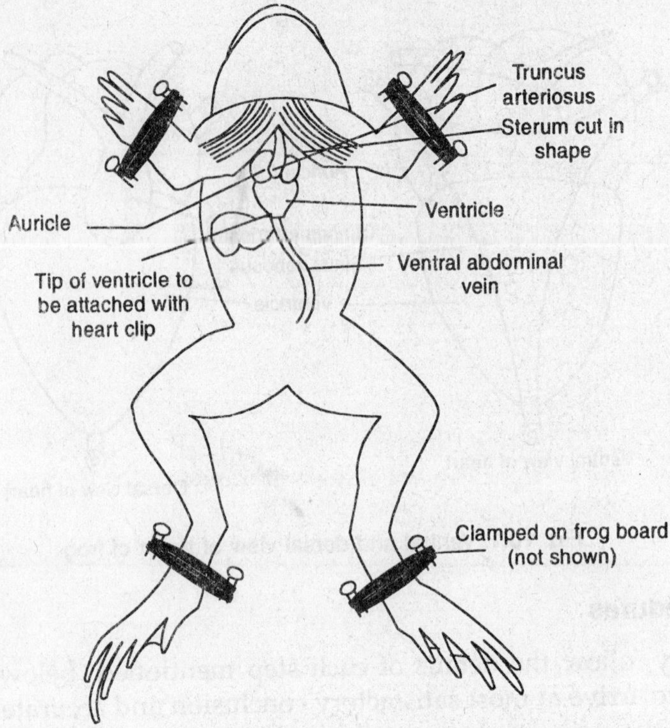

Fig. 10.2. A ^ shaped cut in pectoral girdle to expose heart (Ventral views).

board. Do not allow the heart to dry. To avoid desiccation, pour a few drops of normal salt solution or Ringer's solution over it. The complete kymograph assembly (Figure 10.3) should be kept ready prior to the preparation of heart and fixing of the frog.

(c) Setting up the kymograph assembly

The heart preparations are delicate, only when the recording device offers minimal resistance then in that case a satisfactory record can be obtained. Heart levers writing on smoked kymograph paper are to be preferred over ink writing ones. Only the glossy kymograph paper should be used. Steps involved in setting up of kymograph assembly and recording of heartbeats are as follows:

(i) Level the kymograph as described in earlier chapter.

(ii) Set up the speed either at 12.5mm/sec or 25mm/sec, so that the drum rotates very slowly

(iii) Put the drum carrying smoked paper on the shaft at the desired height and lock it tightly to avoid its slipping during recording.

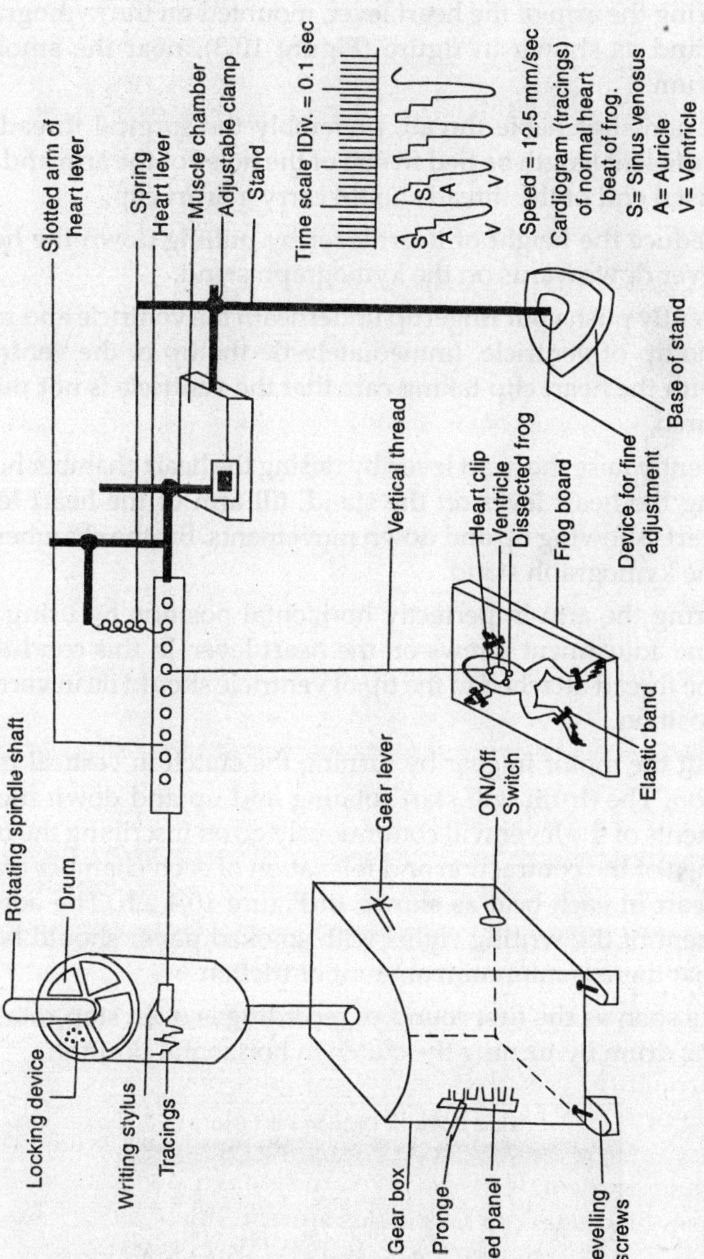

Fig. 10.3. Kymograph assembly for recording the normal heart beat of frog

(iv) Bring the arm of the heart lever, mounted on the kymograph stand as shown in figure (Figure 10.3), near the smoked drum.

(v) A non-stretchable thread, preferably the surgical thread, of sufficient length be tied in one of the holes of the arm and the distal end of the thread should carry a heart clip.

(vi) Reduce the height of heart lever by pulling down the heart lever downwards on the kymograph stand.

(vii) Gently push your fingertip underneath the ventricle and raise the tip of ventricle. Immediately tie the tip of the ventricle with the heart clip taking care that the ventricle is not punctured.

(viii) Gently raise the heart lever, by raising the heart chamber holding the heart lever on the stand, till arm of the heart lever starts showing up and down movements. Fix the chamber on the kymograph stand.

(ix) Bring the arm in perfectly horizontal position by using the fine adjustment screws on the heart lever. In this condition, the thread attached to the tip of ventricle should lie in vertical position.

(x) Put the motor in gear by turning the clutch in vertical position. The drum will start rotating and up and down movements of the lever will continuously go on inscribing the tracings of the contraction and relaxation of each chamber of the heart in each beat as shown in Figure 10.4 a,b. The adjustment of the writing stylus with smoked paper should be so that there is minimum amount of friction.

(xi) As soon as the first round of recording is over, stop rotating the drum by turning the clutch in horizontal position.

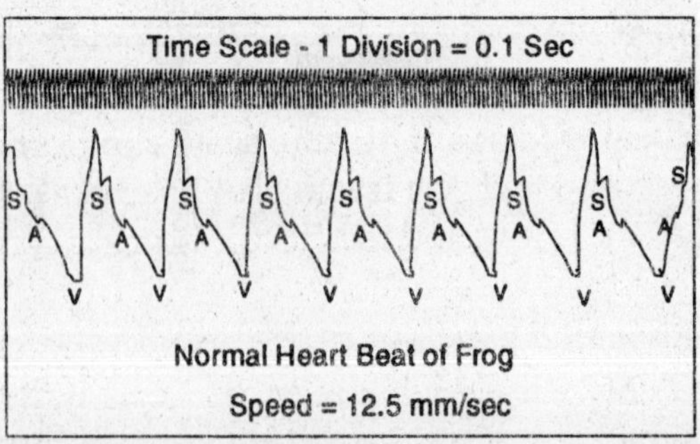

Fig. 10.4a. Cardiogram of normal heart beat of frog at 12.5 mm/sec.

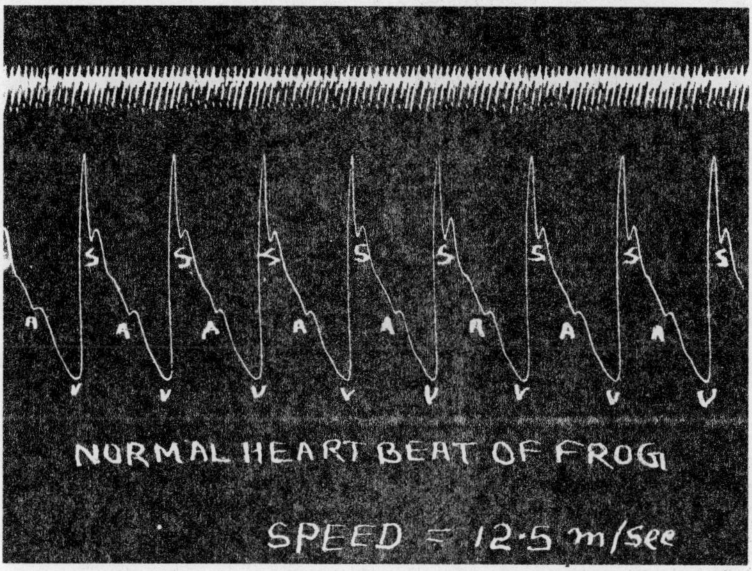

Illustration of original tracing of Fig. 10.4a.

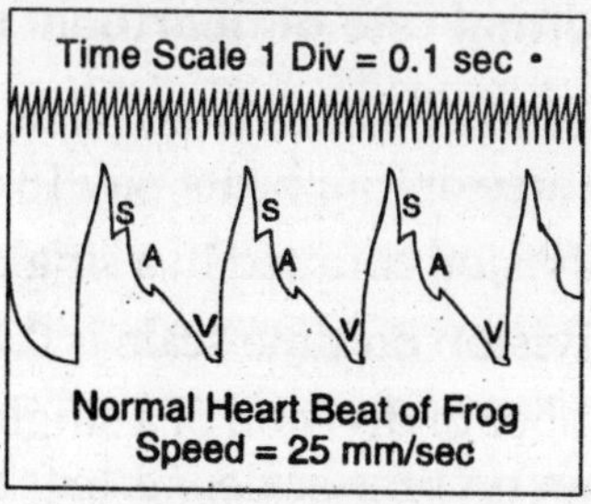

Fig. 10.4b. Cardiogram of the normal heart beat

(xii) Gently push away the writing stylus from the drum.

(xiii) Using an electrical time marker, make a time scale a little above the heartbeat tracings. The time scale should be made at the same speed at which the tracings of heartbeat were recorded. As soon as the one round rotation of drum is complete, switch off the time marker and remove it away from the drum.

The heart should be kept moist all through the experiment by using saline water or Ringer's solution. In no case, the heart should be moistened when the actual recording is in progress.

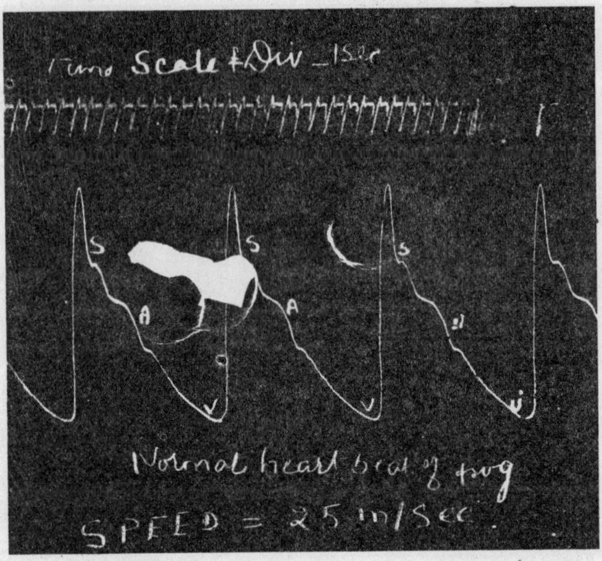

Illustration of original tracing of Fig. 10.4b.

For calculating the normal rate of heartbeat proceed as under.

(1) Count the number of divisions from peak to peak on the time scale covered in three or four beats, say 18 divisions in 3 beats.

(2) The number of division counted in a single beat will then is 6.

(3) The value of 1 division on time scale is 0.1sec.

(4) Multiply the number of division of a single beat with the time value of 1 division on time scale, *i.e.* 6x0.1=0.6sec.

(5) Divide 60 seconds with the time value for a single beat, *i.e.* 60/0.6=100. 100 beats per min is rate of normal heartbeat of frog, which is actually the final result.

Precautions

(1) The heart should be kept moist with Ringer's solution or normal salt solution throughout the exercise, but it should not be moistened during the actual recording.

(2) The speed of the drum should be at medium slow, *i.e.* 12.5 mm/sec, but for greater details of different phases of each chamber medium fast speed, *i.e.* 25mm/sec.

(3) Be sure that the heart lever is counter balanced so that it is not stretched unnecessarily even to a very small extent.

(4) From its normal resting horizontal position, the lever should dip downward step by step with a rising increase at each step in accordance with the chamber of heart concerned

Exercise No. 2

To study the effect of drugs on the normal heart beat of frog.

Apparatus: Kymograph assembly as in exercise no. 1, Ringer's solution plus drugs like adrenaline solution (1:1,00,000) and acetylcholine (=0.01%), urethane solution (20% or 40%). The time scale should be so adjusted that each division = 1.0sec.

Theory: The cardiac muscle is peculiar in the sense that it is much more sensitive to drugs when they are directly applied on them as compared to other muscles. Adrenaline stimulates the sympathetic nervous system because it is a sympathomimetic drug. On the heart, it produces augmentation of contractile forces and acceleration of the heartbeat rate. Acetylcholine, on the other hand, inhibits the activity of heart. Since adrenaline brings about augmentation in heart activity, it is administered in an emergency of failing heart. Similarly there are other drugs, e.g. muscarine, an alkaloid found in rotten betel nuts that slows down or totally inhibits the heart activity; while atropine causes an acceleration of heartbeat by blocking the cholinergic action of vagus nerve.

Procedure

(i) Set the speed of the drum at 25mm/sec.

(ii) Take a few tracings of the normal heartbeat of frog, make a time scale and stop the drum.

(iii) With the help of microdropper pour four or five drops of adrenaline solution on the heart and wait for sometime, till the effect of drug is persistently noticeable.

(iv) Take a few tracings, make a time scale and stop the drum.

(v) Moisten the heart with Ringer's solution till the required conditions of the normal heartbeat of frog is persistently noticeable. Again take a few tracings, make a time scale and stop rotating the drum.

(vi) Pour four or five drops of acetylcholine mixed with Ringer's solution and wait for sometime till the heartbeat become

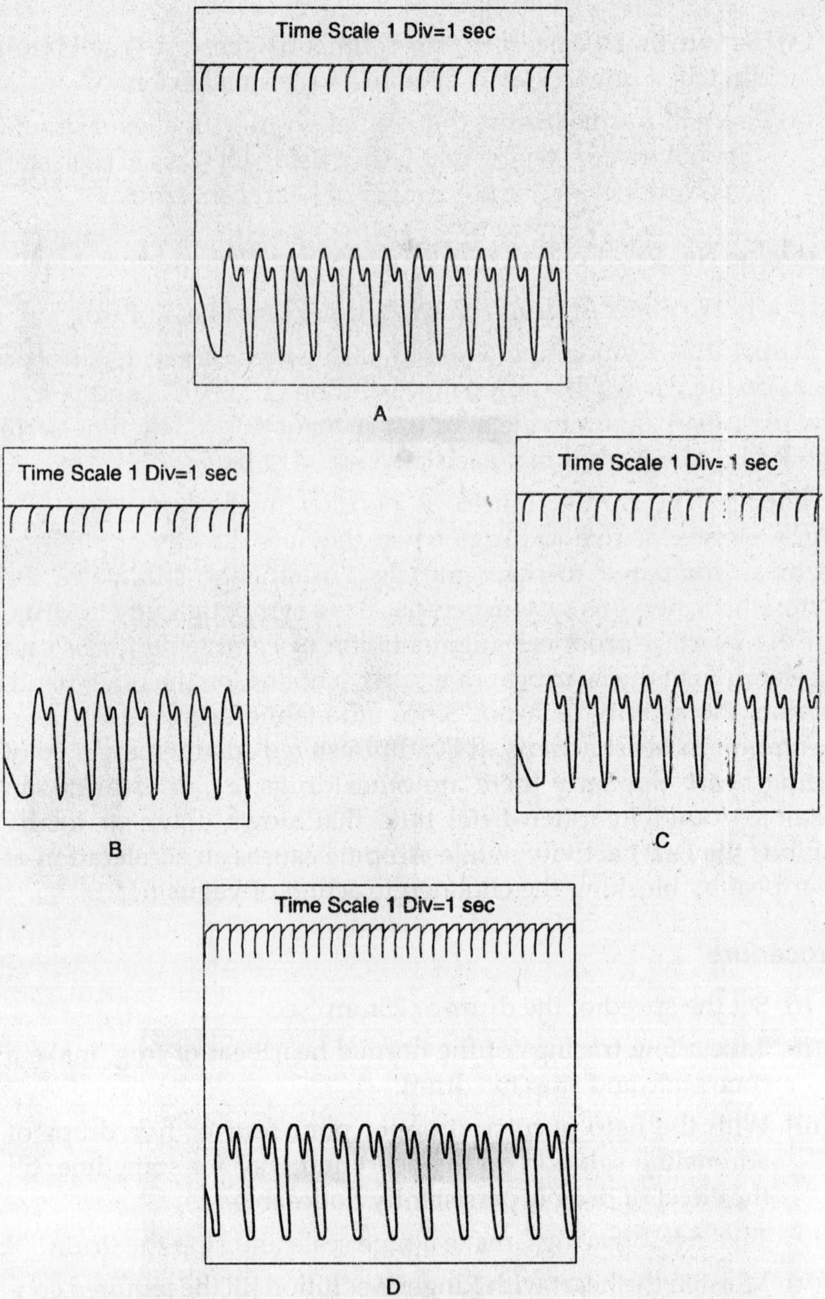

A. Normal heart beat B. Effect of adrenline (enhanced)
C. Regained heart-beat D. Effect of acetylcholine (retarded)

Fig. 10.5. Cardiograms of frog's heart beat showing effect of drugs at speed 25 mm/sec; time scale 1 division = 1 sec.

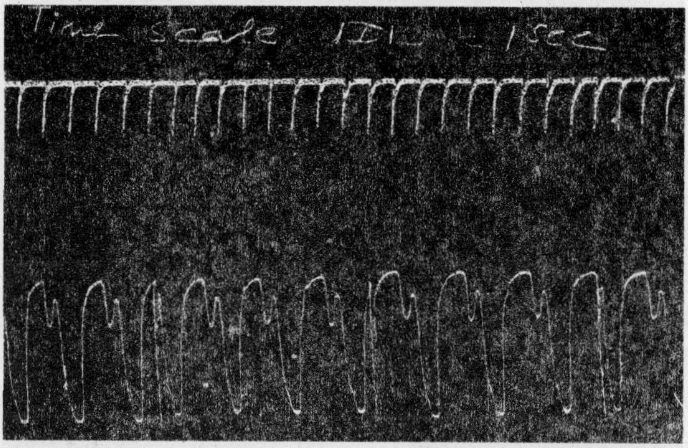

A

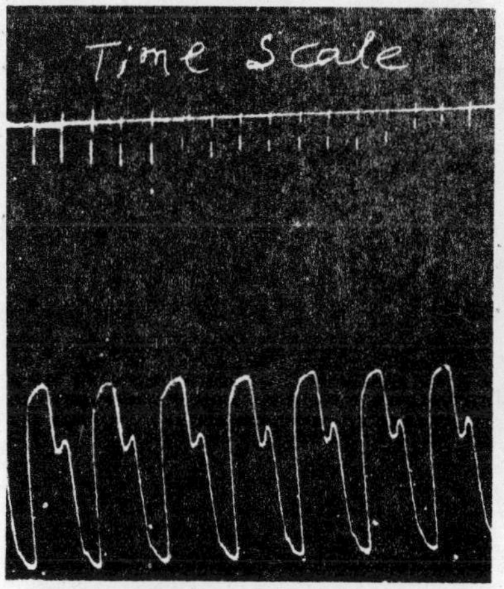

B

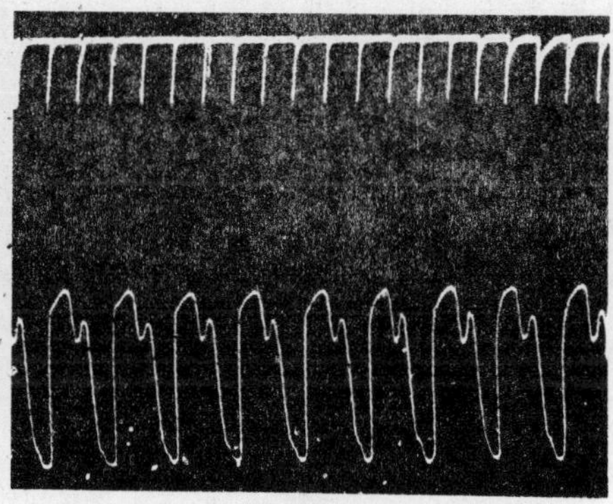

C

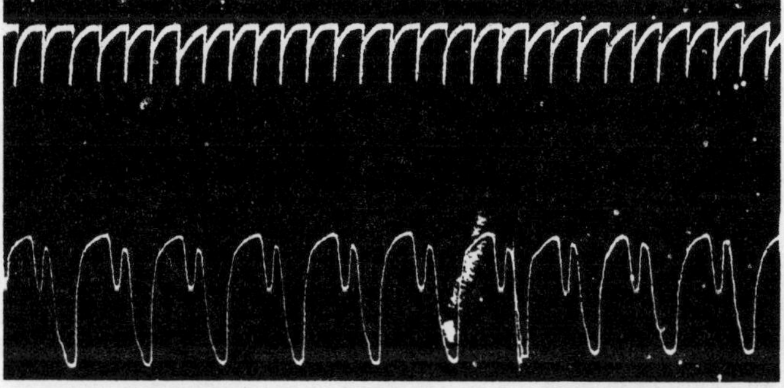

D

Illustration of original tracing of Fig. 10.5.

persistent, *i.e.* becomes almost constant. At this stage, take a few tracings, make a time scale and stop rotating the drum.

(vii) Wash the drug on the same heart with Ringer's solution alone; wait for sometime till the heart regains activity more or less similar to the normal activity. Take a few tracings, make a time scale and stop rotating the drum.

(viii) Take off the drum and remove the paper. Title and label all tracings and make note of the temperature. Finally fix the tracings to make a permanent record.

(ix) Compare your tracings with an illustration of effect of drugs as shown in Figure 10.5.

Adrenaline accelerates the activity of all types of hearts while acetylcholine inhibits the activity of myogenic hearts of vertebrates and molluscs but accelerates the activity of neurogenic hearts, like that of crustaceans, *Limulus*, some insects and annelids.

Observations

(a) In case of adrenaline, note the increase in the amplitude and rate of heartbeat as compared to the normal, Figure 10.5 (b).

S. No.	Condition of heart	Time for single heartbeat	Rate / min	Amplitude
1.	Normal heartbeat at 26°C			
2.	Adrenaline treated heartbeat			
3.	Regained heartbeat after washing with Ringer's solution			

(b) In case of acetylcholine note the decrease in rate of heartbeat and amplitude (Figure 10.5c).

S. No.	Condition of heart	Time for single heartbeat	Rate / min	Amplitude
1.	Normal heartbeat at 26°C			
2.	Acetylcholine treated heartbeat			
3.	Regained heartbeat after washing with Ringer's solution			

Conclusion

Adrenaline is a heartbeat-accelerating drug while acetylcholine is an inhibitory drug, so far as their effect on normal heart beat of frog is concerned.

Exercise 3

To study the effect of electrolytes on the normal heart beat of frog.

Apparatus and reagents: As in exercise no. 1 plus 1% solution of calcium chloride and 1% solution of potassium chloride, Ringer's solution, normal salt solution (for cold blooded animals), cane sugar solution (7.35%).

Theory: All the ions do not affect the heartbeat in the same manner. It is well known that the effect of each individual electrolyte is deleterious but when they are used jointly in the form of a mixture in the solution, they have a beneficial effect.

Procedure

Proceed step by step as under using the same arrangement of kymograph assembly as in exercise no. 1.

(1) Prepare a heart from a freshly narcotized frog and mount the same on a frog board.

(2) Take tracings of a few heartbeats and simultaneously make the time scale. Stop rotating the drum.

(3) Pour one or two drops of 1% calcium chloride solution mixed with sodium chloride solution over exposed heart, wait for 30-60 sec and allow the drum to rotate to make a few tracings of heartbeat. Simultaneously make the time scale. Stop rotating the drum.

(4) Put one or two drops of Ringer's solution. Wait for 30-60sec and start rotating the drum and make a few tracings of heartbeats along with the time scale. Stop rotating the drum.

(5) Pour one or two drops of 1% potassium chloride solution mixed with sodium chloride solution over exposed heart, wait for 30-60 sec. Start rotating the drum and make a few tracings along with the time scale. Stop the drum.

(6) Pour one or two drops of Ringer's solution over exposed heart. Wait for 30-60sec. Start rotating the drum and take a few tracings of heartbeats along with the time scale. Stop rotating the drum. Move away the writing stylus, the lever arm and the time marker from the drum.

(7) Take off the drum from the kymograph and remove the pa-

per. Label all tracings with full details and make a permanent record.

Observation

Note that with each of the electrolyte, *i.e.* calcium chloride and potassium chloride, when applied separately, the rate of heart beat is reduced, *i.e.* the activity of heart is inhibited, but when they are used together in the form of Ringer's solution containing cane sugar or glucose, the rate of heartbeat is increased, *i.e.* it is accelerated.

Exercise 4

To study the effect of ions or salts on the isolated heart of frog.

Theory: Heart continues to beat for many hours outside the animal body in a solution of properly balanced salts.

Material and reagents: Petri dish, a graduated pipette, dissecting instruments, frog, urethane solution (25%), cane sugar solution (7.36%), sodium chloride solution (0.7%), calcium chloride solution (1%) and potassium chloride solution (1%).

Procedure

Anaesthetize a frog preferably by double pithing or by injecting 3ml of 25% urethane solution with a hypodermic syringe into the dorsal lymph spaces under the skin. Expose the heart and count the number of heartbeats in seconds (it should be around 14 beats/ ½ min). Tie a ligature around the aorta, lift up the heart and excise it. To make sure that sinus venosus is also taken out, cut well around the sinus venosus and its vein and proceed as under:

(1) Place the heart in the Petri dish containing 50ml of 7.36% solution of cane sugar (In this condition, the heart is moistened with 0.7% sodium chloride solution) and wait for sometime till heart shows a consistency in beating. Count the number of heartbeats in 30seconds (it should be about 11 beat/½ min.)

(2) Transfer the same heart in another Petri dish containing 50ml of 0.7% sodium chloride solution and wait for sometime and count the number of beats/ ½ min. When the activity has declined significantly then add 0.5ml of 1% solution of calcium chloride to sodium chloride solution and after sometime count the number of beats/ ½min.

(3) When the heart start showing marked deterioration in its ac-

tivity, replace the last solution with 50ml of 0.7% solution plus 0.5ml of 1% KCl solution, wait and count number of beat/ ½ min.

(4) Finally replace the solution of step 3, with Ringer's solution, wait and count number of beat/ ½ min.

Observation

Tabulate your results in the format of table given below and draw conclusion in each case.

S. No.	Name of solution	No. of beat/ ½ min.					Total average result
		1	2	3	4	5	
1.	50ml cane solution 7.36%						
2.	50ml sodium chloride 0.7%						
3.	50ml sodium chloride 0.7% + 0.5ml 1% $CaCl_2$						
4.	50ml sodium chloride 0.7% + 0.5ml 1%KCl						
5.	Ringer's solution						

Results: In general, the heartbeat increases in sodiun. chloride solution and sodium chloride + calcium chloride solution but in sodium chloride and potassium chloride solution and in Ringer's solution, the heartbeat decreases.

(1) The heart continues to beat for many hours outside the animal body in a "balanced salt solution" or "balanced physiological saline", such as Ringer's solution.

(2) The effects of any single element are multifarious depending on the conditions and previous history of heart.

(3) In general, sodium has a stimulating action on heart pacemaker and favours contraction of heart muscles.

(4) In the perfused heart, potassium has a striking depressant action on the contractility and conducting systems. In excess of potassium, a heart undergoes relaxation and stops in diastole.

(5) The action of calcium on heartbeat is predominant and is marked by enhancement of contraction resulting in systole arrest. In the absence of calcium, the heart relaxes and stops in diastole. Calcium is essential for contraction, as it antagonizes the effect of K and Na on the pacemaker and slows down the heartbeat.

In general, sodium shows a stimulating effect, potassium acts as a depressant while calcium appears to be essential for activity of the heart.

Exercise No. 5

To study the effect of temperature on the normal heart beat of frog.

Apparatus: Kymograph assembly as in exercise no. 1, Ringer's solution, crushed ice, thermometer, spirit lamp, beaker, dropper, *etc.*

Theory: Usually the chemical reactions take place with a greater velocity under the warmer conditions (*i.e.* at higher temperatures within an optimal range) rather than at the room temperature. It has been noticed that rise in temperature of 10°C causes the velocity of reaction to double or triple. During heart contractions, lactic acid is formed, which in normal condition is destroyed or removed. By raising temperature, lactic acid gets accumulated in the heart, which stimulates the heart. So applying warm Ringer's solution (for frog) heartbeats are accelerated. If temperature is raised above 45°C, ***Rigor mortis*** occurs due to large accumulation of lactic acid, which causes coagulation of muscle proteins. At this temperature heartbeat stops. Reverse is the case with the cold Ringer's solution, *i.e.* in the cold conditions, the frequency and intensity of heart beat decreases.

Procedure

Set up the speed at 25mm/sec and using the kymograph assembly as in previous exercises, proceed step by step as under.

(i) Expose the heart of a freshly narcotized frog *in-situ* and mount it on a frog board.

(ii) Record a few tracings of the normal heartbeat of frog and make a time scale and stop rotating the drum.

(iii) With the help of dropper or pipette gradually go on pouring drop by drop, the ice cold Ringer's solution (at 0°C or at 4 or 5°C) and wait till the heart shows a consistency in the beating activity. Now, record a few tracings of the heartbeat in this condition make a time scale and stop rotating the drum. Notice a reduction in the rate and slowing down of each individual heartbeat. During recording in this condition, it should be ensured that the temperature remains at 0°C.

(iv) Gradually pour a few drops of Ringer's solution at room temperature over the exposed heart and wait for a few minutes till the heart regains its normal heartbeat to some approximation. Make a few tracings in this regained condition of heart. Make a time scale and stop rotating the drum.

(v) Treat the heart as above with a few drops of Ringer's solution heated to 5°C in a beaker and continue the treatment till it shows consistency in its beat. Take a few tracings of heartbeat at this temperature and make a time scale. In this way, treat the heart at regular intervals of a few minutes with Ringer's solution heated to 10°C, 15°C, 20°C, 25°C, 30°C, 35°C and 40°C. At each temperature, the tracings should be recorded only when during each treatment the heart apparently shows consistency in its contractility. Also make a time scale above the tracings of heartbeat at each temperature level and stop the drum.

(vi) Take off the drum from kymograph, remove the paper from drum and label all tracings with other details. Make a permanent record.

(vii) Tabulate the observation as under.

S. No.	Condition of heart	Rate of heart beat	Observation
1.	Normal heart	beats/min	
2.	Heart treated Ringer's solution at 0°C	beats/min	
3.	Heart treated Ringer's solution at 5°C	beats/min	
4.	Heart treated Ringer's solution at 10°C	beats/min	
5.	Heart treated Ringer's solution at 15°C	beats/min	
6.	Heart treated Ringer's solution at 20°C	beats/min	
7.	Heart treated Ringer's solution at 25°C	beats/min	
8.	Heart treated Ringer's solution at 30°C	beats/min	
9.	Heart treated Ringer's solution at 35°C	beats/min	
10.	Heart treated Ringer's solution at 40°C	beats/min	

The reduction in the rate of heartbeat and slowing of each individual heartbeat on treatment with Ringer's solution at 0°C and a subsequent rise in the rate of heartbeat with successive rise in the temperature upto 30°C or 35°C and thereafter an irregularity in heart beat.

Conclusion: The fall in temperature results in slowing down the contractile activity of heartbeat but successive rise results in increase in heartbeat rate, *i.e.* brings about rapidity of contraction but only upto an optimal temperature after which it brings about impairement of regularity in heartbeat.

Exercise No. 6

To study the role of pacemaker of frog's heart in generating the electrical currents (impulse) at regular interval.

Apparatus: Kymograph assembly as in exercise no. 1 and the surgical thread.

Theory: A crescentric band of specialized muscle cells located at the junction of sinus venosus and auricle is known as pacemaker or sinoauricular (S.A.) Node. S.A. node has intrinsic rate of generating impulses, which are conducted from node to the entire heart, causing the myocardial muscle to contract, thus controlling the rate of the heartbeat. Hence, it is known as the pacemaker. If S.A. node is experimentally destroyed or its activity is blocked, the heart fails to maintain the regulated rhythmicity.

Procedure

To study the above, proceed step by step as under. Adjust the speed at12.5mm/sec.

(1) Expose the heart of a freshly narcotized frog (preferably of bigger size) *in situ* and mount on the frog board.

(2) Insert a ligature below the dorsal aorta and gently lift the heart towards the head and make a stannous ligature round the white crescentric line (corresponding to the pacemaker) that marks the sinoauricular junction and tie exactly over this line. Make a few tracings of heartbeat in this condition and draw the time scale.

(3) Take a few tracings of the heartbeat of frog and make a time scale. Stop rotating the drum.

(4) Take off the drum from kymograph; remove the kymograph paper bearing tracings of heartbeat and time scale. Label all tracings and make a permanent record of tracings.

Observations

Note down the observations and draw conclusions.

Exercise No. 7

Demonstration of the auto-rhythmicity (innate rhythmicity) at regular intervals of the heart of frog.

Apparatus: Petri dish, Ringer's solution, frog.

Theory: Since the rate of the heart activity is under the control of pacemaker and not under nervous control, the isolated heart continues to beat at regular intervals because of the development or origin of the electrical impulses at the sino-auricular node at regular intervals. The exact cause of the generation of pacemaker impulse is not known except that it is a process of rhythmic repetitive oxidation.

Procedure

Narcotize a healthy frog by double pithing, not by using urethane solution for this experiment. Dissect the frog so as to expose the heart. Dissect out the heart by cutting the aorta, sinus venosus and the veins that bring blood to it. Place it in a Petri dish containing Ringer's solution. Note down the number of beats per minute with the help of a timepiece.

Observation and Conclusions: Notice that the heart still beats like the normal one, *i.e.* at regular intervals and this beating continues for a long time. The beating of the isolated heart at regular intervals for a longer duration and the correspondence of the rate of heartbeat with that of normal rate of heartbeat of the heart *in situ* clearly indicates that the rhythmicity of heartbeat is not under the nervous control and regularity in heartbeat indicates that it is controlled by the pacemaker, which is still intact in the isolated heart.

11

Physiology of Muscle Contraction

INTRODUCTION

Most of the mechanical work done by animals is achieved through muscles. Nevertheless, with certain exceptions, like the bone cells (calcified), almost all cells show some movement. The muscle cells, however, exhibit the perfect ability of movement and that to a greater extent because of their physico-chemical architecture, which is oriented towards accomplishing motion solely due to the presence of a protein of contractile nature. The results of architectural elements in muscle cells are simply responsible for the maintenance of the moving parts.

The muscles are broadly distinguished into three types: (i.) *Slow muscles* composed of non-striated or smooth or unstriped or involuntary muscle fibres that are found associated with visceral organs, like the alimentary canal, urinary bladder, *etc.* (ii) *Fast moving* muscles that are composed of striated muscle fibres constituting voluntary muscles, striped muscles or skeletal muscles that are found associated with thigh-bone or hind limb, (iii) *Medium-fast muscles* referred to as cardiac muscles associated with the heart.

The architecture of each muscle fibre is so designed as to facilitate alternate contraction and relaxation. Muscles have two characteristic mechanical properties, namely the contractility and elasticity.

When any muscle is stimulated by any type of stimulus, may be physical, chemical, electrical, *etc.*, it undergoes different types of contractions. The various phenomena are differentially named, such as single muscle twitch, staircase or treppe, fatigue and tetanus.

In this chapter, the exercises have been designed to study the various phenomena and intricacies involved in them with the help of gastronemius nerve muscle preparation from a freshly narcotized frog. This chapter comprises of three parts as under.

Part A deals with the setting up of kymograph assembly for recording the muscular contractions. **Part B** deals with the preparations of a gastronemius nerve muscle preparation of frog. **Part C** deals with experimental exercises to study various phenomena under varying conditions, strength, intensity and duration of the stimulus applied for.

PART A

Setting up of Nerve Muscle Experimental Assembly for exercises based on muscle contractility exhibiting various phenomena of muscular contractions.

Requirements: Kymograph instrument, recording drum, solid-state electronic stimulator. Frog muscle chamber with clamp and drum, pair of adjustable silver electrodes, adjustable muscle lever (after load lever with weights), filaments, kymograph adjustable stand with quick micrometric tangential adjustment provided at the base of the stand for to and fro movement of the vertical stainless upright.

Procedure

(i) Place the kymograph on a glass plate kept on horizontal top of the working table and level it in a perfectly horizontal condition by using a spirit level and operating the levelling screws.

(ii) Adjust on the left side, the desired speed on the speed panel according to the nature of experiment.

(iii) See on the right side that the clutch is in horizontal position ensuring that the motor is not in gear as well as the on and off switch should be in off position.

(iv) Connect the plug of kymograph with the main AC and make all electrical connection with drum on one hand and electrodes on the other to bring them in circuit as and when required.

(v) The student stimulator is nowadays very frequently used in laboratories as it has virtually replaced the induction coils, vibrating reeds, variable interruplers, various keys and low voltage supply units of batteries.

Adjustments to keep the drum in circuit: This is made possible when the drum is connected to the drum terminals of the stimulator and the short-circuiting key is closed. Connect the 'stim-out' terminals to the electrodes. Keep the Pulses/ Drum in switch on 'Pulses' position. Switch on the mains. Start rotating the drum. Gradually increase the voltage on the voltage dial till with the drum you get response. This is the minimal stimulus. Such electrical connection and adjustments are needed for exercises like the tetanus.

To keep the drum out of circuit: This is done when the drum contacts are connected to drum terminals of the stimulator, simply by opening the short-circuiting key. The frequency switch should not be on the drum position. Connect the 'stim-out' terminals to the electrodes. Manipulate the frequency knob to the required frequencies. Put the Pulses/ Drum switch on 'Pulses' position. Increase voltage gradually till you get response.

To get single impulses: Keep the Pulses /Drum switch on 'Drum' position. Short the drum terminals. Connect the 'stim-out' terminals to the electrodes and switch on the mains. Press the push button to get a single impulse. Control the voltage to the desired strength.

Precautions

(i) Do not short the output terminals.

(ii) Do not press the press keys, while drum is in circuit.

(iii) For getting a single impulse, press the key for a moment and release.

PART B

Preparation of the gastronemius nerve muscle preparation.

(a) ***Dissection of frog for the preparation of nerve muscle complex.*** Proceed step by step as under:

(i) Narcotize a frog by either injecting 2 or 3 ml of 20% or 40% urethane solution (depending upon the size of frog) with a hypodermic syringe (5ml) in the dorsal lymph sinuses under the skin or by double pithing.

(ii) Make a transverse circular cut around the middle of the belly of frog. Hold the frog with duster in one hand and with other hand hold very tightly the skin of lower half of belly and pull it down towards the legs in a single determined pull. This process helps to peel off the skin rather quickly and largely.

(iii) Lay the frog in a dissecting tray with ventral side facing upwards and cut open the belly, remove the visceral organs and expose the sacral nerve complex. Cut the vertebral column transversely with a bone cutter just above the sacral nerve complex (Figure 11.1a).

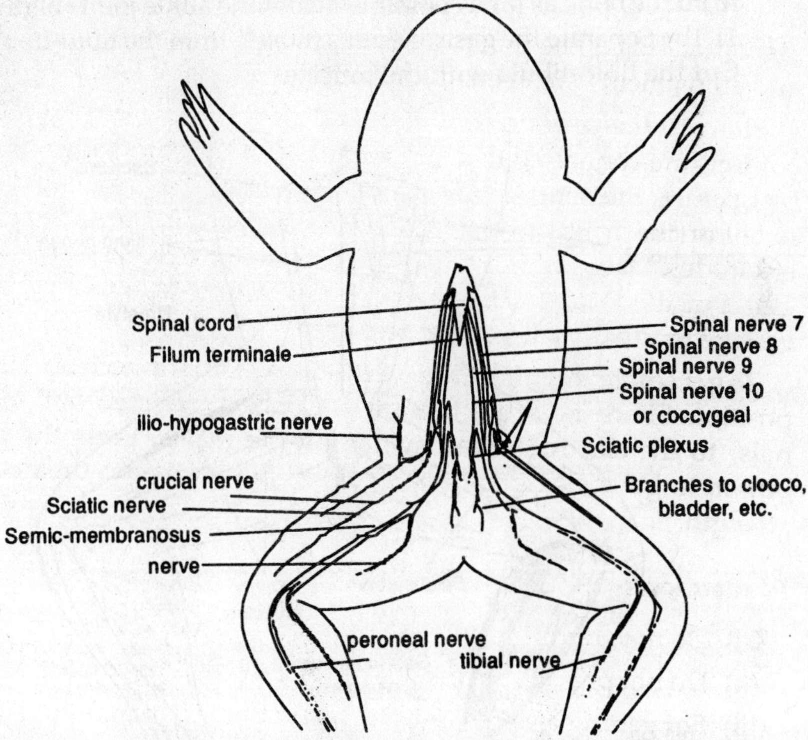

Fig. 11.1a. Dissected frog showing the origin and course of sciatic nerve

(iv) Lay the frog in the tray on its ventral surface. Grasp the urostyle with the help of a stout forceps and cut away the muscle from its external lateral sides right upto its base. Finally remove the urostyle by cutting it with a bone cutter at the level of its base.

(v) Cut the connection of ilium with the transverse processes of 9th vertebra and bisect the portion of vertebral column.

(vi) Hold the piece of vertebrae attached to the three roots of sciatic nerve with the help of forceps, trace the yellowish white sciatic nerve down towards thigh and by exposing the muscular mass on the thigh dissect it down further upto the knee. See the sciatic artery side by side, while tracing the nerve down towards the knee. Go on cutting the smaller branches. The sciatic nerve innervates the gastronemius muscle, which is to be used in all exercises of muscle physiology in next Part C.

(vii) Cut through the tendon Achilles at the down end of muscle from the bone as far as possible below the ankle joint (Figure-11.1b). Separate the gastronemius muscle from the tibio-fibula. Cut the tibio-fibula with the muscle.

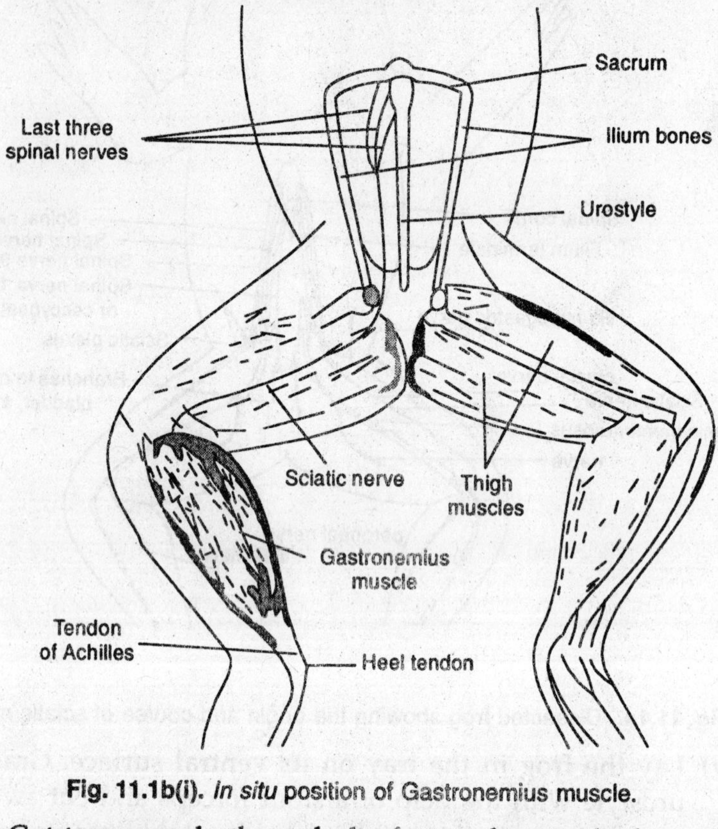

Fig. 11.1b(i). *In situ* position of Gastronemius muscle.

(viii) Cut transversely through the femur close to the knee. Now the preparation (Figure-11.1c) is ready for experiment. Transfer it in a large Petri dish and keep it moist with occasional pouring of one or two drops of saline water over the preparation.

During dissection avoid pulling or stretching of the nerve and do not apply any pressure on nerve, otherwise preparation will be rendered quite useless for recording the various contraction phenomena. Nothing stands in the way of success in this preparation except lack of dissection skill in operation. Avoid contact of nerve with frog's skin. Do not touch nerve with hand or instruments.

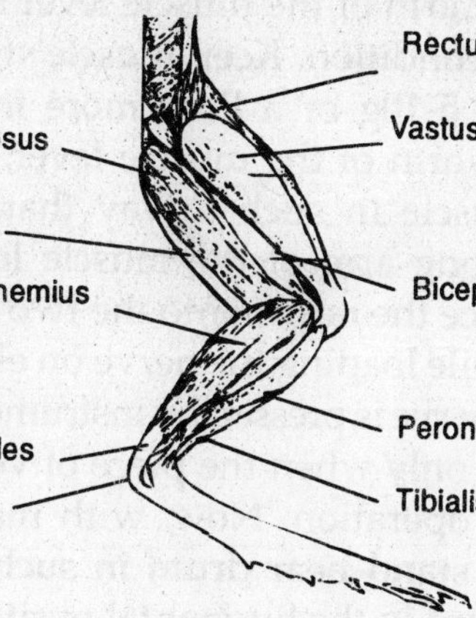

Fig. 11.1b(ii). *In situ* position of Gastronemius nerve muscle preparation of frog.

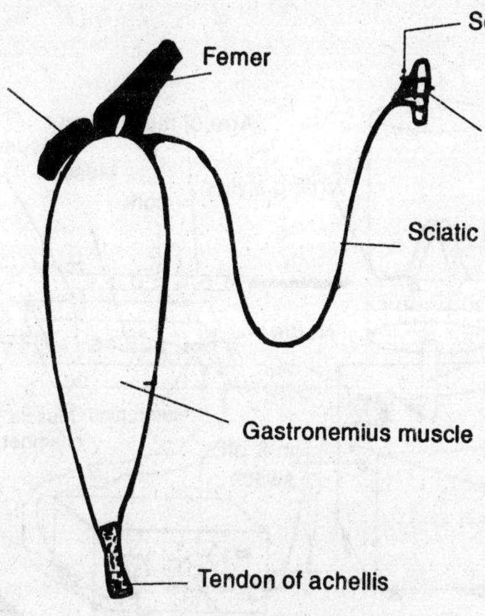

Fig. 11.1c. Finally prepared gastronemius nerve muscle preparations of frog.

(b) ***Mounting of nerve muscle preparation in muscle chamber:*** Tie a non-elastic surgical thread around the tendon Achilles. Hold the thread with one hand and the other piece of vertebral column wit forceps and transfer the preparation in muscle chamber without stretching the nerve. Fasten tightly the knee joint with pin on the cork fitted at left hand side in a hole at the bottom of chamber. Make a loop of thread fasten to tendon of Achilles and attach it to the hook of the short arm of the muscle lever so that the muscle is just in the stretched condition. Keep muscle stretched by suspending small weight of 5-10g or a little more in the weight holder attached to the long arm of the muscle lever, as shown in Figure 11.2. Adjust the muscle in such a way that its contraction, *i.e.* stimulation of the long arm of the muscle lever always goes or moves upwards. Place the nerve onto the two terminal ends of the silver electrodes. While loading the nerve on electrodes takes every care that neither the nerve is pressed by instrument nor it is stretched. This can be avoided only when the piece of vertebrae is held with forceps during this operation. Now, with manual manipulation, shift the myograph stand near drum in such a position that the long lever arm remains in the horizontal position when the muscle is in resting position and the writing stylus touches the drum at a tangent to the surface of the drum. Let the writing stylus touch the smoked drum very lightly without much friction during its movement.

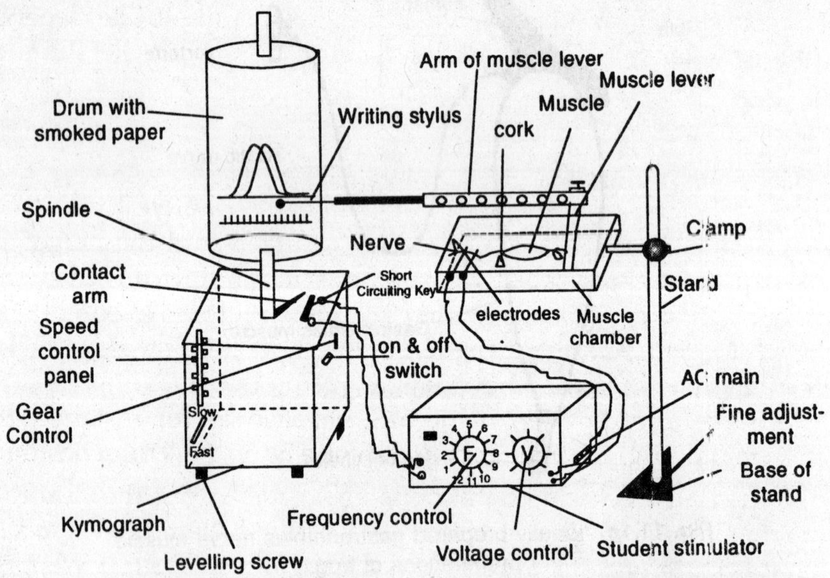

Fig. 11.2. Kymograph assembly for experiments on muscle contraction.

PART C

Experimental exercises to study the various phenomena of muscle contraction

Exercise No. 1

To record a single muscle twitch in the gastronemius nerve-muscle preparation of frog.

Theory: If the muscle is subjected to a single transitory stimulus, it reacts with a momentary contraction followed by relaxation. This is known as a single muscle twitch. The record of tracing on kymograph paper is called ***myogram***. When the muscle is stimulated, at first there appears no apparent change but after a short while the muscle contracts. The period of gap between the point of stimulation and emergence of response is called the latent period or simply latency. The latent period is followed by a phase of contraction, which is followed by a phase of relaxation in which muscle again regains its original length and shape. In frog nerve-muscle preparation, the duration of latent period varies from 0.01 to 0.02 second. The duration of phase of contraction varies from 0.04 to 0.05 seconds and that of the relaxation phase from 0.05 to 0.06 seconds, whereas the total time for a single muscle twitch may vary from 0.09 to 0.11 sec. The height of the amplitude depends upon the length of the muscle and the intensity of the stimulus applied.

Procedure

For this exercise set the speed at 320mm/sec. Before loading the nerve-muscle preparation in muscle chamber and the nerves on electrodes, make sure that the short circuiting device is in such a position that the drum is not in circuit and it comes in circuit only when the drum arm or arm presses the short circuiting key during the rotation when all other adjustments on the student stimulator dial have already been made and checked as for instance, the selector knob on the frequency dial should be on the 'Drum position'.

When all electrical connections are made to put the drum, as well as the electrodes in circuit and verified them, then proceed ahead to perform the experiments with the freshly prepared nerve muscle preparation. In each ensuing experiment, you have to be cautious about the position of the Pulses/ Drum. Switch on the dial of the stimulator for final recording. Now, proceed further step-by-step as follows:

(i) Keep "on and off" switch of kymograph in "off" position and the clutch in the horizontal position.

(ii) Set the speed on the speed panel at 320mm/sec.

(iii) Adjust the myograph in such a position so that the long arm of the lever remains in a horizontal position and this along with writing stylus remains at a tangent to the surface of the drum.

(iv) The recording should always be done over/above the lower margin of drum.

(v) Let the writing stylus touch the smoked paper very lightly. Make the base line by manually rotating the drum for one complete rotation. This should always be drawn slightly away from the seam or joint of the paper.

(vi) The contact drum should be so adjusted that on coming in contact with the short circuiting key in open position, the recording is made somewhere near middle of the smoked paper. When such an adjustment has been made, then with the drum arm in contact with short circuiting key, make a point of stimulation by just flickering over the writing stylus on the base line.

(vii) Rotate the drum anticlockwise so that the writing stylus comes to be on the left side of the mark of point of stimulus on the base line.

(viii) Put the drum switch in "on" position and simply bring the clutch in vertical position just for one complete rotation of the drum. As soon as one complete rotation is over, myogram of a single muscle twitch has been recorded. Stop the rotation of the drum by bringing switch in horizontal position.

(ix) Remove away the writing stylus of lever.

(x) Bring the writing stylus of time marker, a tuning fork (Figure-11.2a) or an electrical time marker, mounted on a separate stand and make a time scale with a time marker, a tuning fork or an electrical time marker, below the base line just by setting the arms of fork in vibration with a starter and allow the drum to rotate at 320 mm/sec speed for only one revolution.

(xi) Bring the main switch of kymograph in "off" position to stop rotating the drum so also the switch of the stimulator.

(xii) Even when the recording is over, try to keep the nerve muscle preparation, so that it may be used for other experiments.

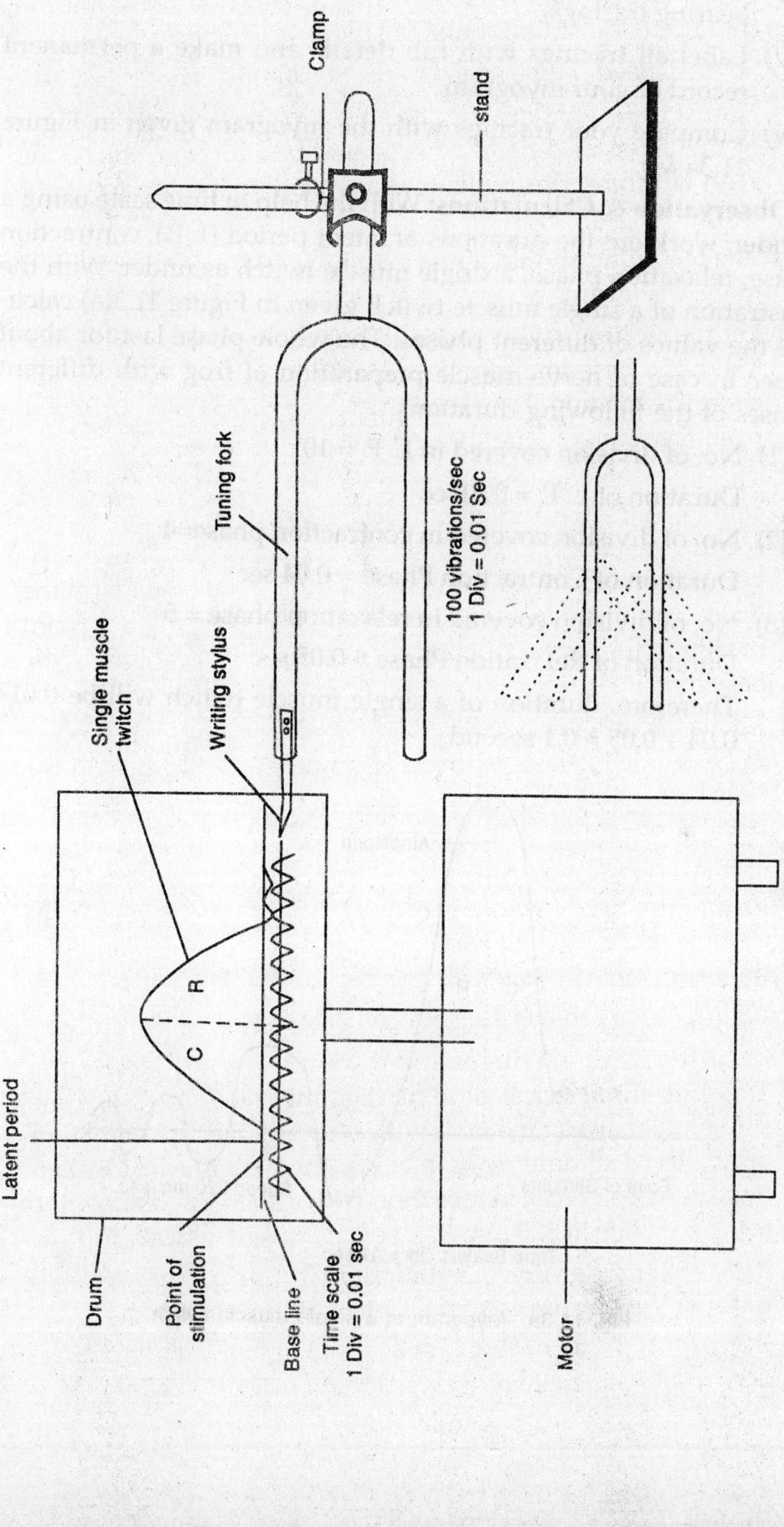

Fig. 11.2a. Diagram illustrating the use of tuning fork as time marker (speed 320 mm/sec.)

(xiii) Take off the drum from kymograph and remove the paper bearing tracings.

(xiv) Label all tracings with full details and make a permanent record of your myogram.

(xv) Compare your tracings with the myogram given in Figure 11.3a &b.

Observation & Calculations: With the help of time scale using a divider, work out the durations of latent period (L.P.), contraction phase, relaxation phase, a single muscle twitch as under. With the illustration of a single muscle twitch given in Figure 11.3(a) calculate the values of different phases. The whole phase last for about 0.1sec in case of nerve-muscle preparation of frog with different phases of the following durations.

(1) No. of division covered in L. P. = 10

Duration of L. P. = 0.01sec

(2) No. of division covered in contraction phase=4

Duration of Contraction Phase = 0.04 sec

(3) No. of division covered in relaxation phase = 5

Duration of Relaxation Phase = 0.05 sec

Therefore, duration of a single muscle twitch will be 0.01+ 0.04 + 0.05 = 0.1 second.

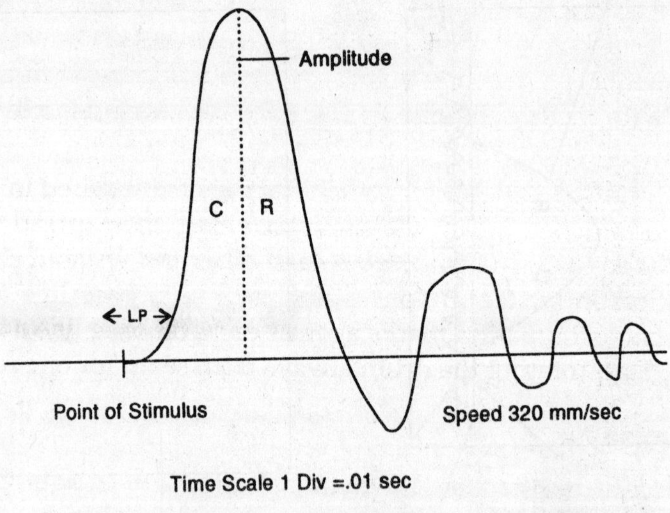

Fig. 11.3a. Myogram of a single muscle twitch

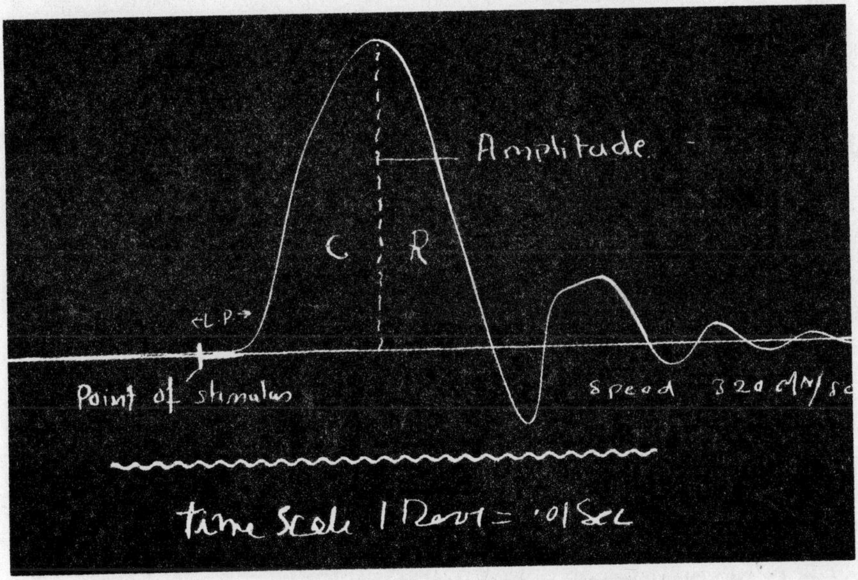

Illustration of original tracing Fig. 11.3a.

Also directly find out the total number of division of time scale covered in a single twitch from the time of emergence of response and finishing of relaxation phase, thereby the duration of a single muscle twitch. To work out the precision in performing your experiment compare the two values obtained alone with each other and with those mentioned earlier for the single muscle twitch of frog.

If more than one single twitch tracings are required in one revolution of drum, *i.e.* on the same base line, then spread the drum arm at equitable distance from each other and without putting the kymograph switch in "on" position draw a base line and make points of stimulation marks against each, on base line as in earlier case. Start rotating the drum always from seam for one revolution.

Exercise No. 2

To record the tracings for demonstrating the treppe or staircase phenomenon using nerve muscle preparation of frog.

Apparatus: Use the same assembly, electrical connections and adjustments as in exercise no. 1.

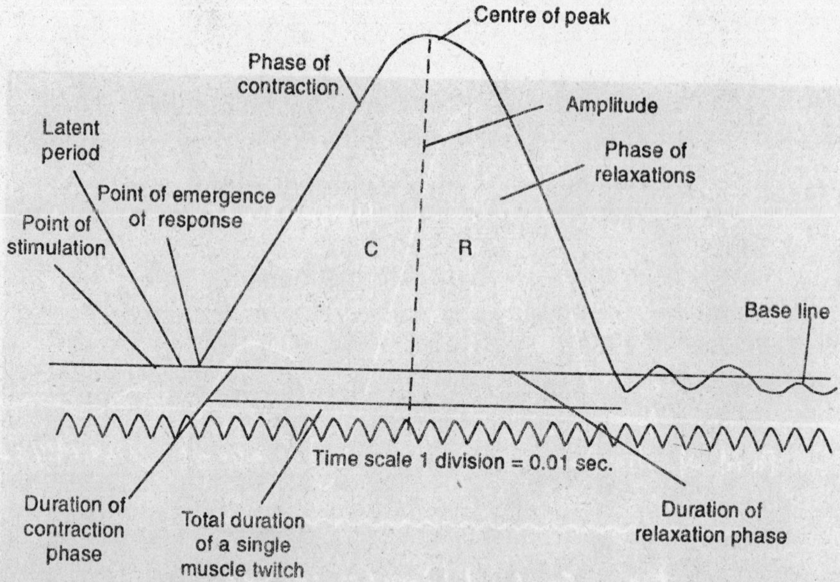

Fig. 11.3b. Diagrammatic representation of the myogram of a single muscle twitch in case of the gastronemius nerve muscle preparation of frog.

Theory: When a muscle is repeatedly stimulated at regular intervals, each time with the same strength of stimulus, the first few contractions increase progressively in amplitude and all phases became shorter. This is called the ***treppe*** or ***staircase phenomenon.*** This is due to the beneficial effect on the contraction and can be explained as under:

(i) Fine branches of nerves innervate only some muscle fibres. These form a motor unit. That means each muscle consists of several motor units. When first stimulus is applied, then only a few motor units get stimulated and only these respond to the stimulus. But, later by other successive stimuli, all the muscle fibre motor units gradually get excited producing maximum response and it is for this reason that a gradual increase in the amplitude is clearly perceptible in record of first few tracings.

(ii) During first few contractions, some internal changes do occur, e.g. rise in temperature, accumulation of CO_2 and lactic acid. All these have beneficial effects on the succeeding contractions.

Procedure

For perfect tracings of this phenomenon proceed step by step as under:

(i) Prepare the nerve muscle preparation and mount in the muscle chamber as for single muscle twitch in previous exercise.

(ii) Draw a base line and mark the point of stimulation.

(iii) Open the short-circuiting key.

(iv) Check that all the electrical connections are intact and the speed, the selector knob of Pulses/ Drum. Switch on the frequency dial to drum position.

(v) Switch on the main switch and allow the drum to rotate for at least three or four complete revolutions at the speed of 320mm/sec (fast speed).

(vi) Stop rotating the drum and remove the writing stylus away from the smoked paper.

(vii) Using a tuning fork (time marker), make a time scale below the base line.

(viii) Take off the drum from the kymograph, remove the paper, label all tracings with complete information regarding speed, value of each division of time scale, *etc.* and make a permanent record of tracings.

(ix) Compare your tracings with myogram in Figure 11.4.

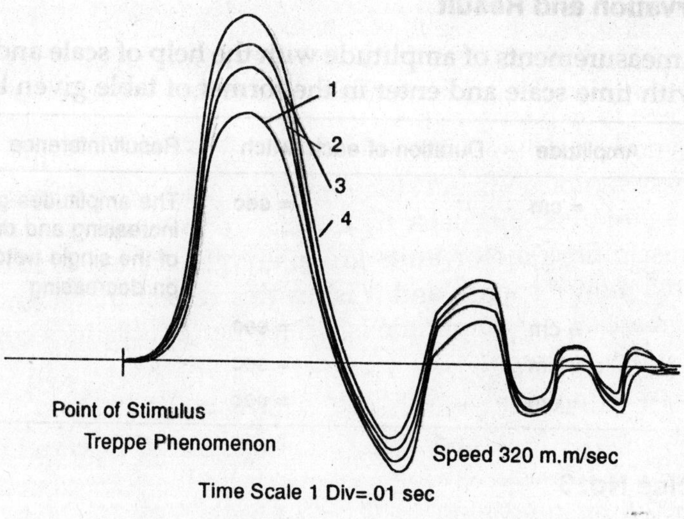

Fig. 11.4. Myogram of Treppe phenomenon

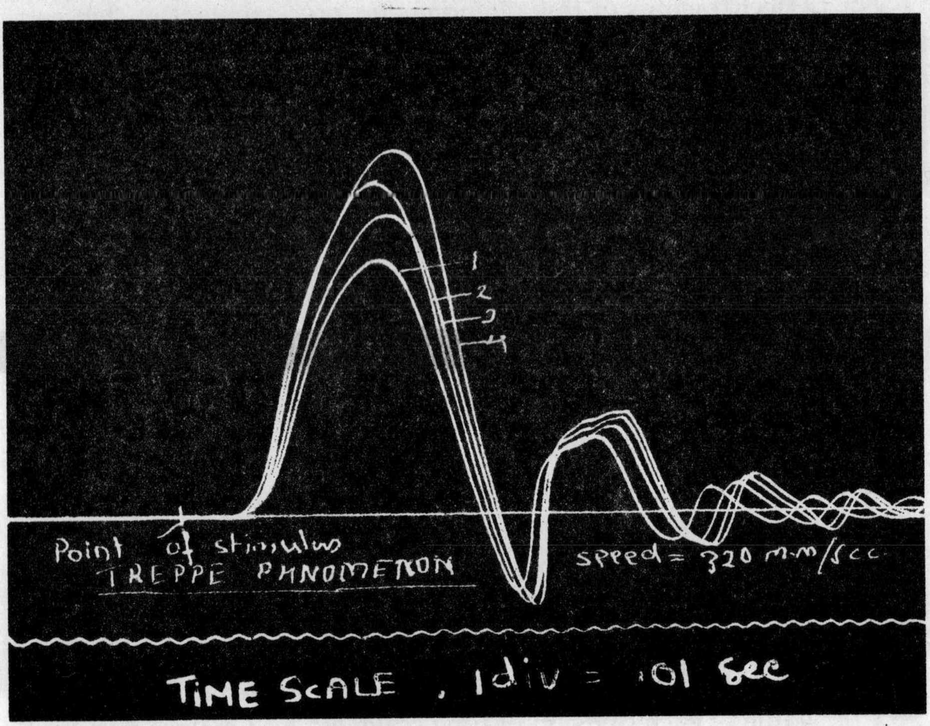

Illustration of original tracing of Fig. 11.4.

Observation and Result

Make measurements of amplitude with the help of scale and duration with time scale and enter in the format of table given below.

S. No.	Amplitude	Duration of each twitch	Result/Inference
1.	= cm	= sec	The amplitudes go on increasing and duration of the single twitch go on decreasing
2.	= cm	= sec	
3.	= cm	= sec	
4.	= cm	= sec	

Exercise No. 3

To record the tracings for demonstrating the genesis of Tetanus in the nerve muscle preparation of frog.

Apparatus: All as in exercise no. 1, except for the following:

(i) The short-circuiting key should be closed.

(ii) The speed on speed panel should be set at 25mm/sec.

(iii) Keep the Pulses/ Drum switch on 'Pulse Position'.

(iv) Put the selector knob on the frequency dial to Drum position.

Theory: Tetanus is a sustained state of muscular contraction and occurs when more than two stimuli are given in rapid succession. When the rate of stimulation is slow, the muscle relaxes partially between each stimulation and the individual impulses or contractions are identifiable and discernible. This is termed as ***incomplete tetanus*** or ***tonus***. However, when stimuli are given at sufficiently short interval, it will result in toothed tetanus. When stimuli follow each other very rapidly, the muscle contracts but are unable to relax completely in between two successive stimuli, hence a complete fusion of contraction occurs resulting in a straight line on the record of tracings. This is called ***complete tetanus.***

Procedure

Before starting the experiment, thoroughly check all electrical connections and then proceed step by step as under:

(i) Set the kymograph assembly and stimulator with all connections intact as in exercise no. 1, except that the short-circuiting key is kept closed.

(ii) Mount the freshly prepared nerve muscle preparation as in earlier exercises.

(iii) Start rotating the drum from the seam (paper joint) at a speed of 25mm/sec.

(iv) Put the frequency knob at 1click per second (cps) and Pulses/ Drum switch on 'Pulses' position. Put the stimulator in "on" condition with mains.

(v) Press the push button to get a few impulses, release the push button and quickly change the frequency knob at 2cps and press the push button to take a few tracings and release the push button. In this way go on recording the tracings with gradual increase in frequencies till the Tetanus invoking condition is arrived. Release the press button and stop rotating the drum.

(vi) Take off the drum and remove the paper, label the tracings accordingly, *i.e.* below each specific recording, put down 1cps,

2cps, *etc.* as well as nature of the tetanized condition of tracings. Make a permanent record of tracings.

(vii) Compare your tracings with the myogram of tetanus given in Figure 11.5 a & b.

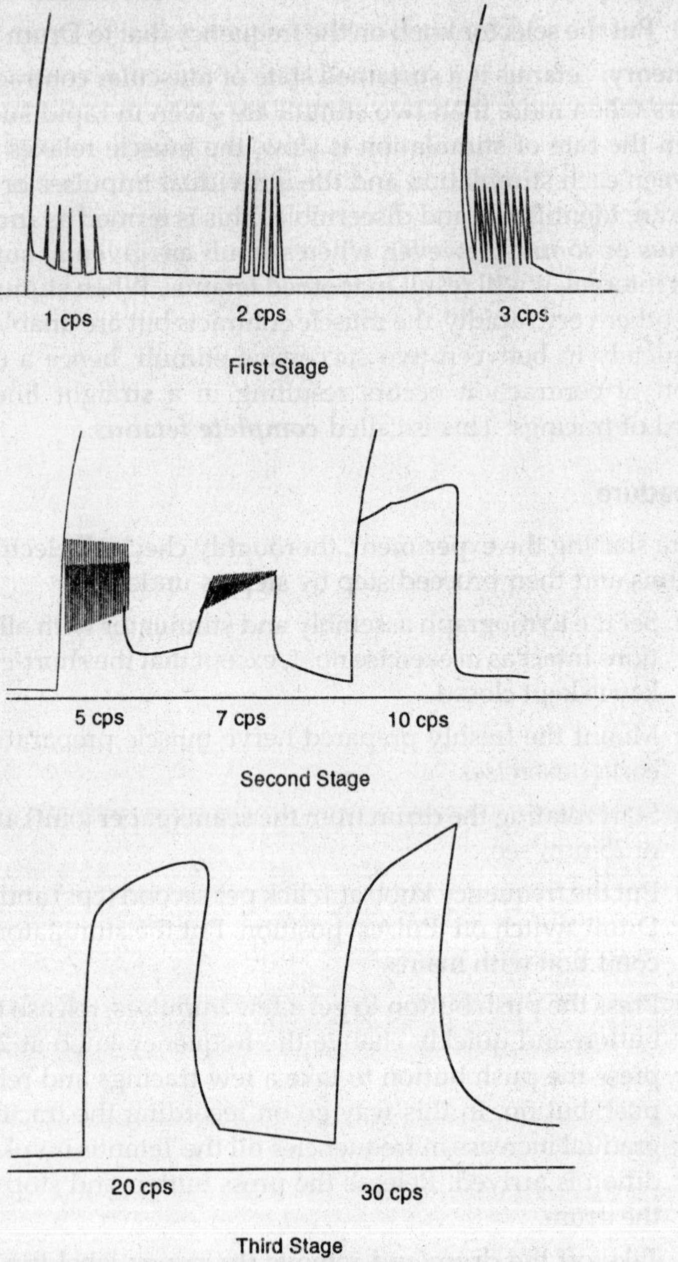

Fig. 11.5a. Myogram of the genesis of tetanus in nerve muscle preparation of a large frog.

First stage

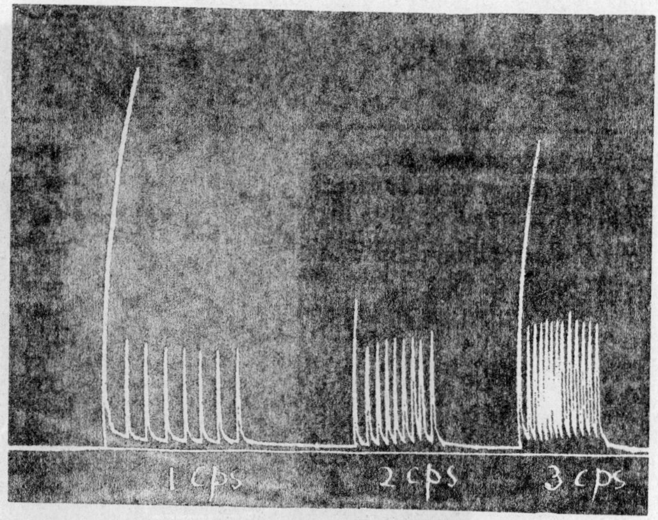

Second stage

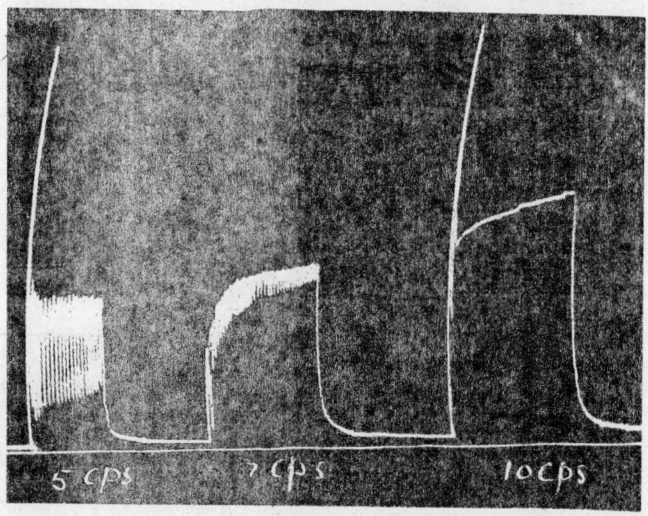

Illustration of original tracings of Fig. 11.5a. (Contd.)

Third stage

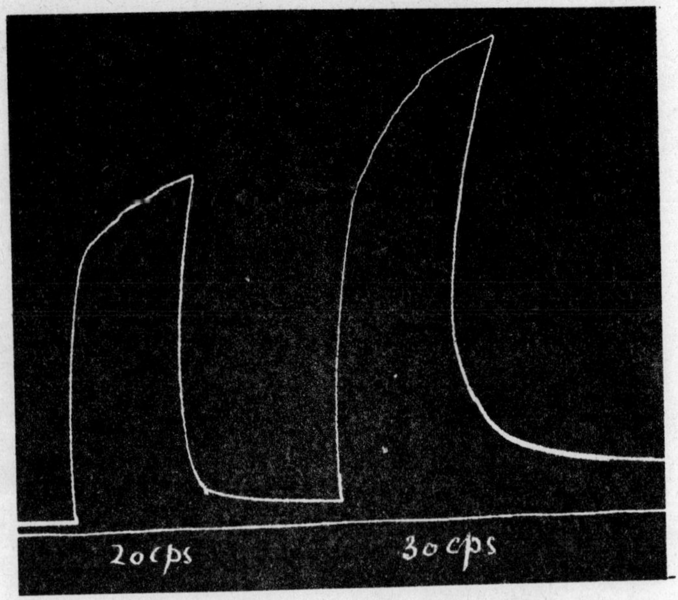

Fig. 11.5a. (Contd.)

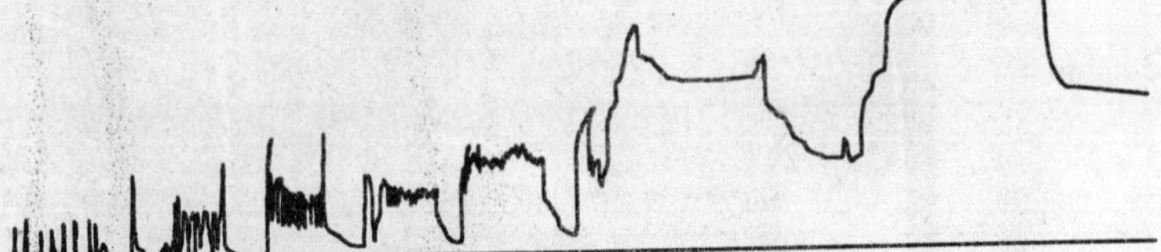

Fig. 11.5b. Myogram of the genesis of tetanus in nerve muscle preparation of a medium sized frog.

Observations and Conclusions: Note that the curve of Tetanic contraction of a muscle under load is considerably higher than that of single muscle twitch and its height goes on increasing with the increase in strength and frequency of stimulation.

Exercise No. 4

Recording of tracings to ascertain the seat of fatigue in a gastronemius nerve muscle preparation of frog.

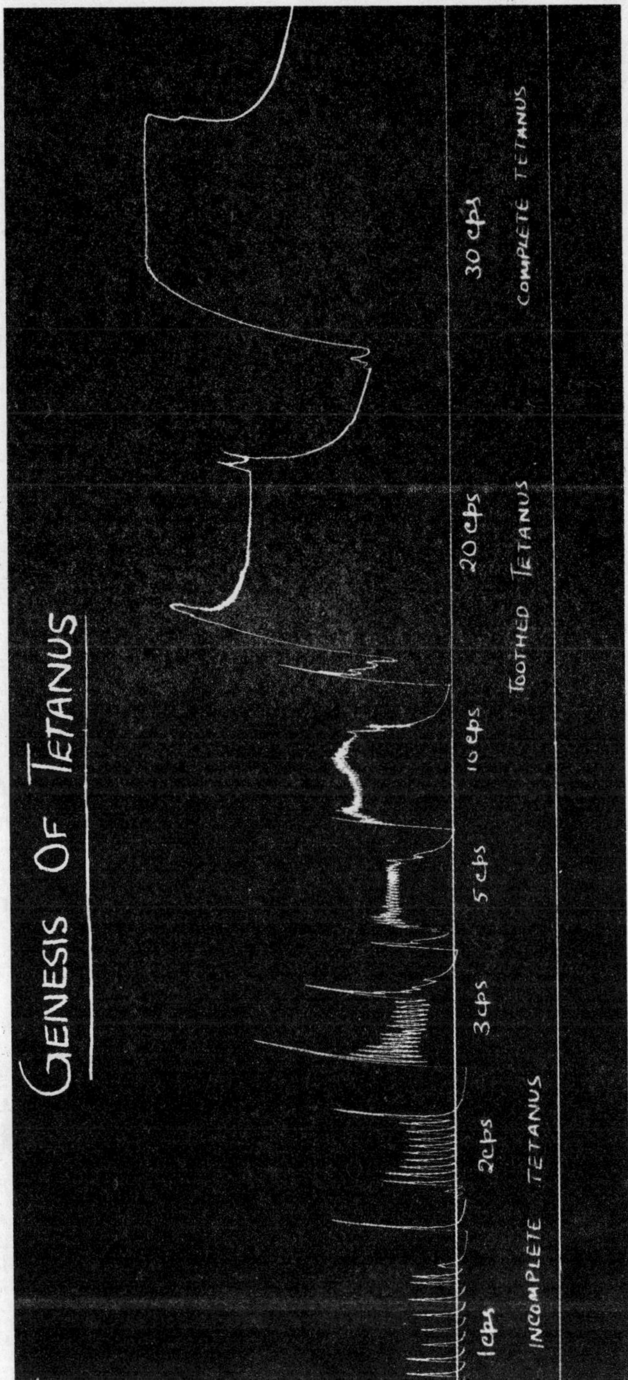

Illustration of original tracing of Fig. 11.5b.

Apparatus: As in exercise no. 1

Theory: When a muscle is stimulated through a nerve by successive stimuli, a state of fatigue is eventually developed and the muscle no longer responds to any more stimulus.

The possible sites of fatigue in a preparation like this are the nerve, the myoneural junction, which exhibits the properties of a synapse and the muscle itself.

Since the muscle (after being fatigued by the stimulation through nerve) on being directly stimulated responds to the stimuli and the nerve being almost infatigueable, the most likely site of fatigue seems to be myoneural junction.

Procedure

Develop fatigue by either of the method given below:

Procedure A

Set the apparatus and connections as in exercise No. 1.

(i) Adjust the speed of the drum on a speed panel at 0.12 or 0.1 mm/sec and keep the short-circuiting key closed.

(ii) Adjust the frequency knob at 1cps.

(iii) Allow the drum to rotate and put on the stimulator in "on" position till the muscle fails to respond, *i.e.* there is no movement in the muscle lever.

(iv) Stop the drum and place the electrodes directly on muscle and apply stimulus.

Procedure B

Set the apparatus and connections as in exercise No. 1.

(i) Adjust drum speed at 320mm/sec and the short-circuiting key open.

(ii) Put the stimulator in "on" condition and start rotating the drum continuously till the muscle fails to respond.

(iii) Stop the drum, and place the electrodes directly on muscle and make a separate tracing.

(iv) Stop rotating the drum, disconnect the stimulator, remove away the writing stylus and take off the drum. Label all tracings and make a permanent record of tracings.

(v) An illustration of the myogram of fatigue by this method is given in Figure 11.6a & b.

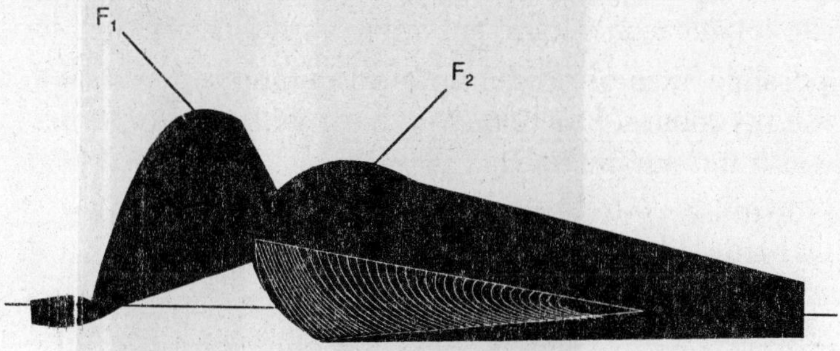

Fatigue by contracture in muscle stimulated through nerve

F_1 = Fatigue developed in muscle by stimulation through nerve

F_2 = Fatigue developed in same muscle by direct stimulation

Fig. 11.6a. Myogram of fatigue

Observations and Inference

See that in method A, the muscle stopped responding to a continuous stimulus that was applied through nerve via myoneural junction but responded when the same muscle was directly stimulated. Thus this shows that the seat of fatigue is myoneural junction (Figure 11.7). So is the case in method B.

Special instructions for the flawless operation of the Student Stimulator

This stimulator is so designed as to enable the students to use the same with full confidence and without any risk in performing the experiments on animal physiology. Another salient feature of this type of stimulator is that it has replaced to a very great extent the Induction coils, Vibrating Reeds, Variable Interrupters, various keys and low voltage supply units or batteries.

Major specifications of this type of stimulator include

(a) Pulse shape: Rectangular

(b) Pulse width: 2m sec (fixed)

(c) Frequency: 1,2,3,5,7,10,20,30,50 and 100 cps.±5%

(d) Output: 0-8V continuous variable with built in key in the drum circuit.

It is easily used on 220V, 50 cycles AC Mains.

Illustration of original tracing of Fig. 11.6a.

Fig. 11.6b. Fatigue developed in the muscle by stimulation through the nerve.

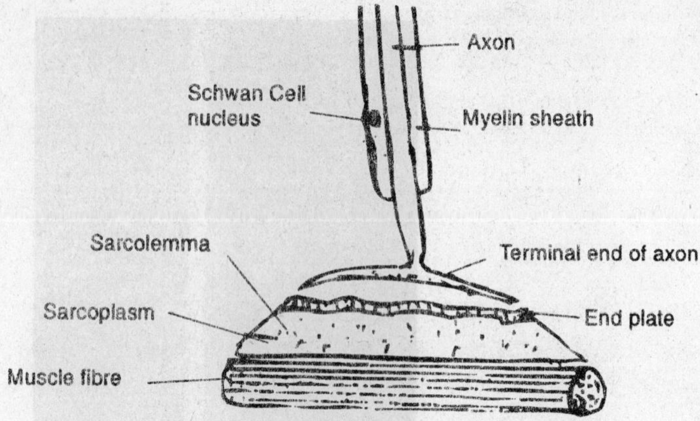

Fig. 11.7. Illustration of myoneural junction

Instructions for a clear conception of its use see the circuit diagram in Figure 11.8.

A. When the drum is required in circuit.

(1) Connect the drum contacts to the 'Drum' terminals of the stimulator. Connect the 'Stim out' terminals to the electrodes meant for stimulation.

(2) Keep the Pulses/Drum switch on 'Pulses' position

(3) Put the selector knob on the Frequency dial to 'Drum' position.

(4) Switch on the main switch. Start the drum. Gradually go on increasing the voltage on the voltage dial till with the drum contact closing you get a response. This will correspond to the minimal stimulus required to arouse a response.

(5) For obtaining the maximal response, go on increasing the strength of the pulse, by increasing the voltage in steps of 0.5V and continue taking tracings till no more increase in the response is noticeable. This will correspond to the maximal strength.

B. When the drum is not required in circuit.

It may be noted that even when the drum contacts are connected to the Drum terminals of the stimulator is not on the 'Drum' position

(1) Connect the 'Stim out' terminals to the electrodes.

(2) Adjust the frequency knob at the required frequency.

(3) Put the Pulses/Drum switch on 'Pulses' position till you get the response.

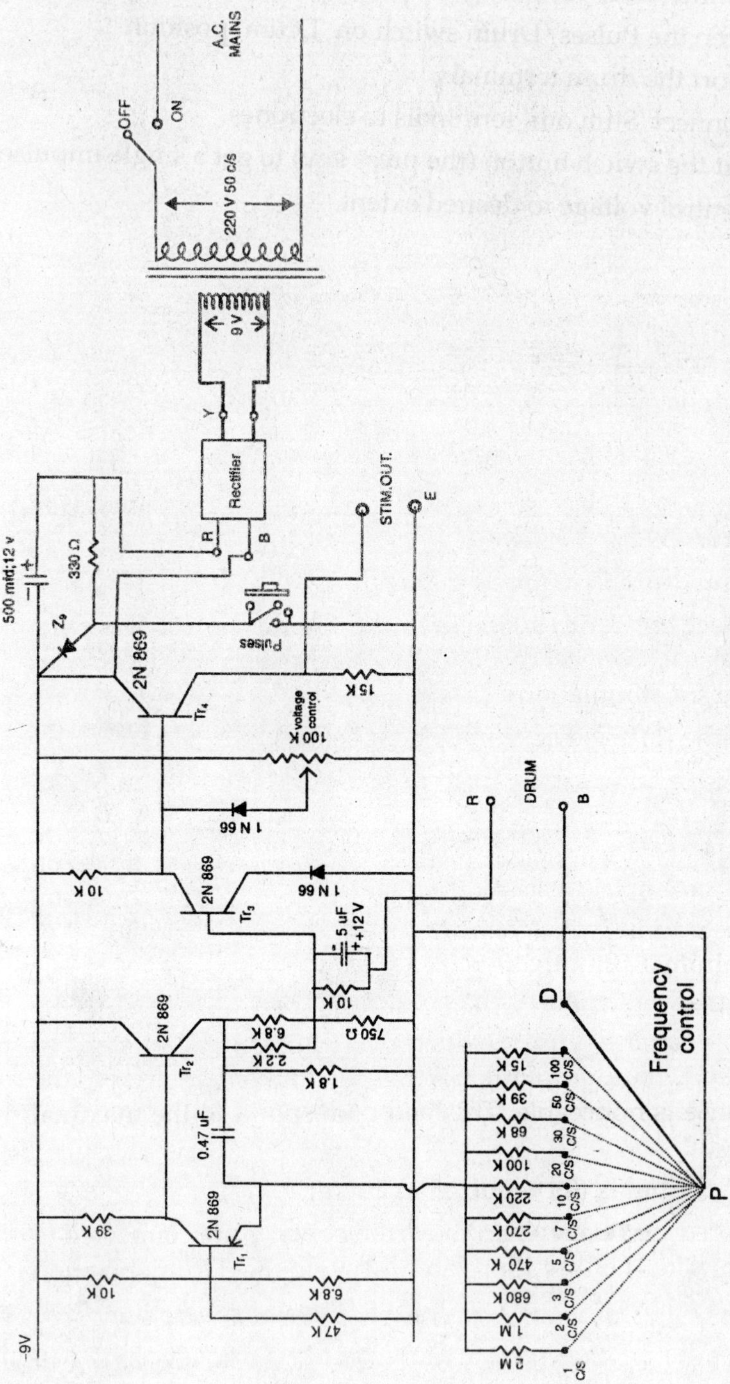

Fig. 11.8. Circuit diagram of student stimulator

C. Adjustments for recording simple impulses.

(1) Keep the Pulses/Drum switch on 'Drum' position

(2) Short the drum terminals.

(3) Connect 'Stim out' terminals to electrodes.

(4) Put the switch button (the press key) to get a single impulse.

(5) Control voltage to desired extent.

12

Physiology of Nerve

INTRODUCTION

The nervous tissue is often referred to as the master tissue of the body for the simple reason that it controls the activity of almost all other tissues of the body. Basically, the nervous tissue is composed of nerve cells called neurons. A typical neuron (Figure 12.1) consists of a cell body, containing a nucleus enclosed in cytoplasm. Each cell body has several short branches called dendrites and a single long axon. These neurons constitute the nervous tissue, which is grey in colour, henc referred to as ***grey matter***. In majority of the cases, the axon is covered with myelin sheath, which is white and usually referred to as ***white matter***.

Nerve impulses pass from one cell to another through the contact of dendrites and the contact zones are referred to as ***synapses***. Nerves are composed of numerous neurons covered by a connective tissue sheath, the myelin sheath. Each neuron is functionally distinguished into three regions, namely the (1) generator region, which is restricted to dendrites, soma and collaterals, (2) the contractile regions, which is confined to axon, and (3) the transmissional region, which comprises the nerve endings or terminals.

Exercise No. 1

Determination of the velocity or the rate of nerve impulse conduction in a gastronemius nerve muscle preparation of frog.

Theory: The conduction of the action potential (*i.e.* nerve impulse) from one place to another along the length of a nerve fibre takes place by local circuits. The local circuits originate from the non-stimulated or positively charged area to the stimulated negatively charged portions. As each new portion is depolarized, it acts as a sink toward which current flows from the adjacent area. Depolarization occurs progressively through the entire length of axon. The true source of the development of current is the flow of $Na+$ ions into the cell. When a stimulus is applied at a particular point on the nerve fibre or nerve, it generates an impulse, which is in the form of a brief electronegative charge created on the surface of the axon (Figure 12.2). The impulse once has been generated, travels along the axon or nerve signal is propagated. The brief electrical negativity is called ***action potential.*** Since it moves along the length of the fibre it is also called a ***propagated action potential.***

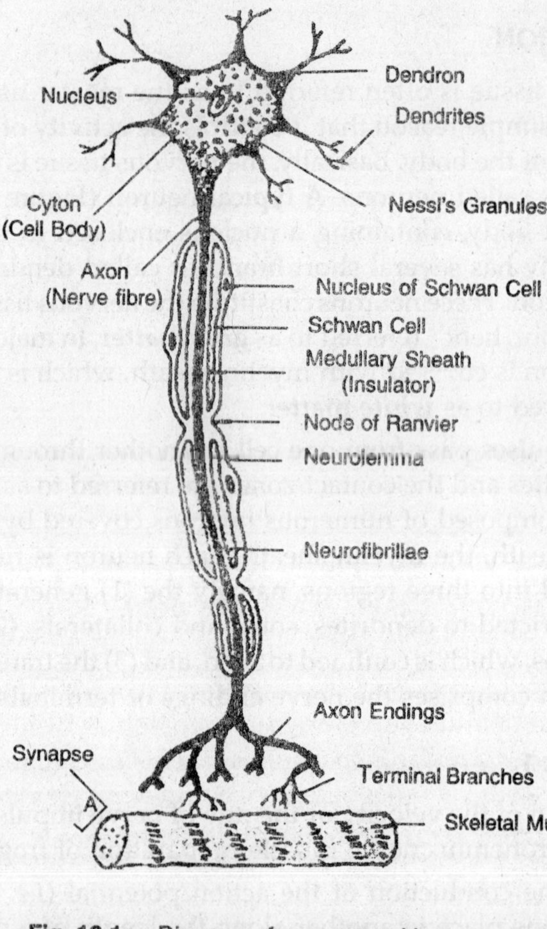

Fig. 12.1a. Diagram of a neuron with synapse

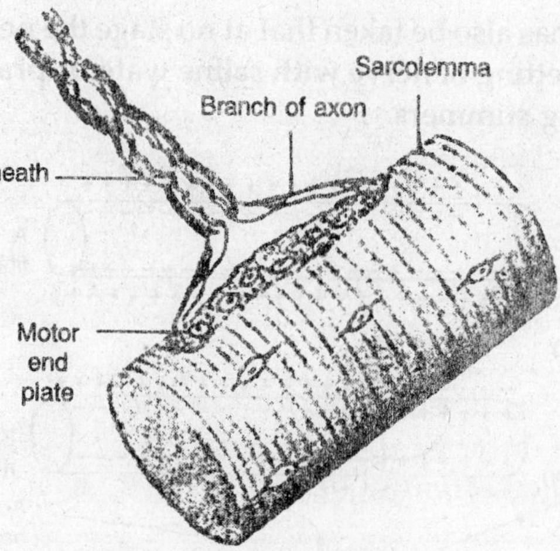

Fig. 12.1b. Diagram of neuromuscular junction.

Once an impulse has been initiated, it travels along the axon at a fairly rapid rate. The rate at which the impulse travels is known as the velocity of conduction. This rate of impulse transmission mostly depends upon the diameter of the fibre and the thickness of myelin sheath. The conduction velocity of the mammalian fibre, which is the largest, is as high as 100 metre/sec; in fibres of smaller diameter about 10 metre/sec and in fibres of smaller diameter about 2 metre/sec. The rate of conduction is the same regardless of the fact whether the fibres are used from squids, crustaceans, frogs or cats or any other mammal in the laboratory experiments.

Apparatus, accessories and reagents required

Kymograph apparatus, muscle lever, muscle chamber, induction coil with wires, key and two dry battery cells, student stimulator, tuning fork, *etc.* and normal saline (for frog).

Procedure

Make a gastronemius nerve muscle preparation from the freshly narcotized large sized frog as in experiments in chapter 11, taking care that the sciatic nerve is taken out completely, right from its exit point from the vertebral column to its point of attachment with muscles. In this experiment great care is to be exercised so that during dissection nerve is not injured or stimulated. All this depends on manual skill of a student. During preparation and experimentation care has also be taken that at no stage the nerve becomes dry. Frequent wetting of nerve with saline water is practically very necessary during summers.

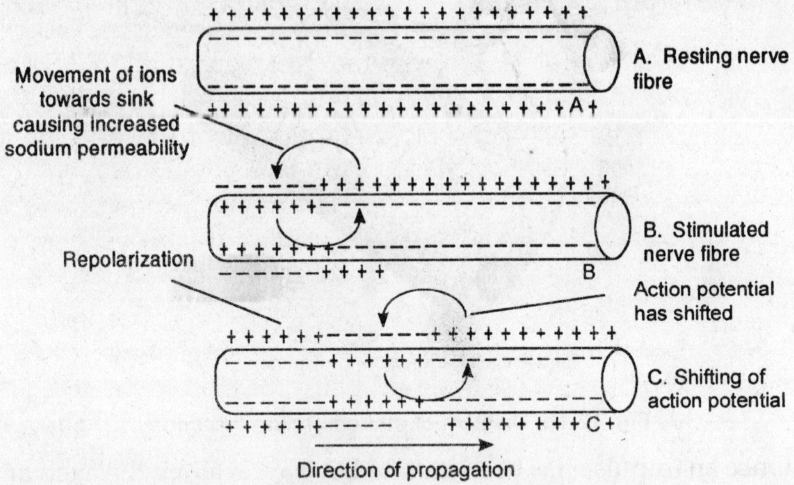

Fig. 12.2a. Illustration of local circuit theory of action potential propagation through unmyelinated nerve fibre

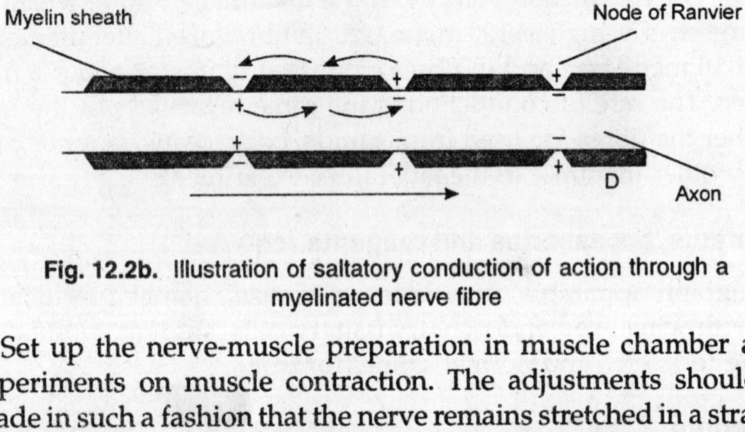

Fig. 12.2b. Illustration of saltatory conduction of action through a myelinated nerve fibre

Set up the nerve-muscle preparation in muscle chamber as in experiments on muscle contraction. The adjustments should be made in such a fashion that the nerve remains stretched in a straight way without any stress, when connected with knob of the muscle lever.

The velocity of nerve impulse can be measured either by using an induction coil or student stimulator. Since the induction coil apparatus is like that of du Bois Reymond Induction Coil based on the use of Galvanic current to stimulate the tissue are no longer in use, hence only the use of Student Stimulator designed for A.C. Current is being used in the ensuing experiments also.

Velocity measurement using student stimulator

Procedure

Prepare the gastronemius nerve muscle preparation, taking all precautions and load the same in the muscle chamber as in earlier case. The nerve should remain in the stretched condition. Set the speed of the drum at the top, *i.e.* 320mm/sec and bring the short-circuiting key in the main circuit only when pressed with the contact arm of the stimulator. At this point, it should be noted that with every thing in on condition that is the stimulator is connected to mains, off and on switch in on position, frequency knock at pulse, the nerve will be stimulated only when the contact arm of the kymograph presses the short-circuiting key.

After having made all connections as for a single muscle twitch, put the stimulator in off position, place the tips of the two electrodes as far away as possible from the neuromuscular junction, *i.e.* at point A of the nerve as in Figure 12.3. Connect the writing stylus with surface of smoked paper and rotate the drum manually, stop it at the point where the contact arm comes in contact with short-circuiting key in open conditions. Simply flick the writing stylus to

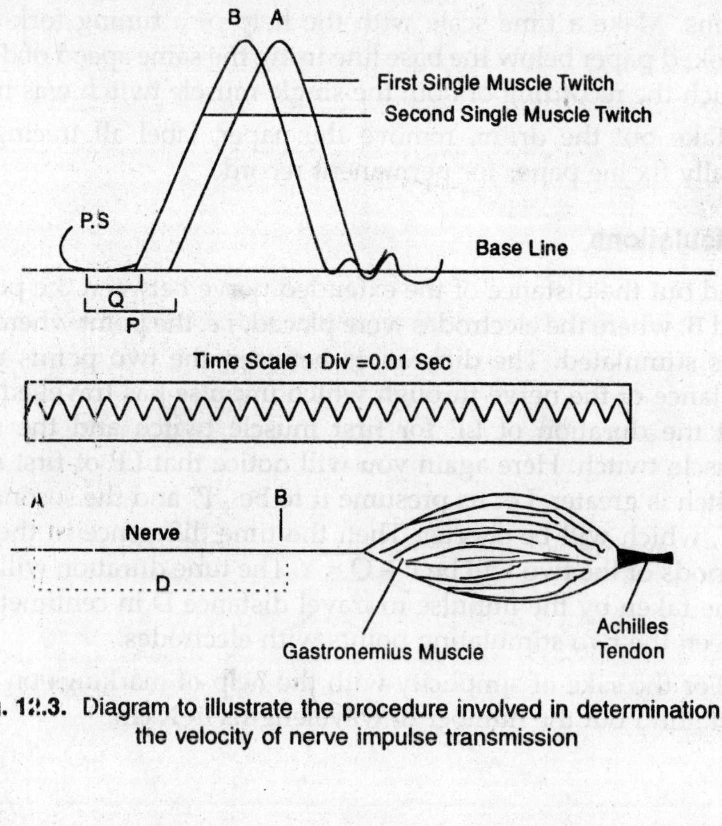

Fig. 12.3. Diagram to illustrate the procedure involved in determination of the velocity of nerve impulse transmission

leave a marking for the point of stimulation. This has to be done simultaneously when the base line is being drawn. Rotate the drum for one complete revolution to make the base line.

Now put the stimulator in on position and rotate the drum for only once by putting it in gear. A tracing of the single muscle twitch will be recorded on the drum. It should be noted that all the contact arms should coincide with each other so that in one revolution of the drum, the short-circuiting key is put in main circuit only for a single stimulus, *i.e.* with short-circuiting key in open position.

As soon as a single muscle twitch with electrodes placed at a point A of the nerve is made, disconnect the stimulator with mains. Place the two tips of the electrodes on the nerve as close to the muscle as possible, *i.e.* nearer to the neuromuscular junction, *i.e.* at point B and put the switch of stimulator in on position. Allow the drum to rotate for only one revolution with all settings as in previous case. This time, again a tracing of a single muscle twitch will be recorded on the smoked paper. The response of the second muscle twitch will be much closer to the point of stimulation, *i.e.* in this case the value of the latent period will be less. Remove the writing stylus away from the drum and disconnect the stimulator with mains. Make a time scale with the help of a tuning fork on the smoked paper below the base line using the same speed of drum at which the recording of both the single muscle twitch was made.

Take out the drum, remove the paper, label all tracings and finally fix the paper for permanent record.

Calculations

Find out the distance of the extended nerve between the points A and B, where the electrodes were placed, *i.e.* the point where nerve was stimulated. The differences between the two points will be distance of the nerve through which impulse has travelled. Work out the duration of LP for first muscle twitch and the second muscle twitch. Here again you will notice that LP of first muscle twitch is greater. Let us presume it to be 'P' and the second to be 'Q', which will be shorter. Then the time difference in the latent periods of the two will be P – Q = Y. The time duration will be the time taken by the impulse to travel distance D in centimetres between the two stimulating points with electrodes.

For the sake of simplicity with the help of markings on tuning fork, find out the number of wavelengths in X cm.

Then the time taken by the impulse to cover the distance between P – Q will be equivalent to

$$\frac{10Y}{X} \times \frac{1}{100} \text{ sec}$$

During this time, the impulse travelled D cm

Consequently in, $\left(\frac{10Y}{X} \times \frac{1}{100}\right)$seconds

the impulse travelled $= \frac{D \times 1}{100}$ of a metre

Then the velocity of nerve impulse at T^0C

$$= \frac{DX}{10Y} \text{ metres per second}$$

Precautions

(1) The nerve muscle preparation should be just kept moist with Frog's Ringer solution throughout the experiment.

(2) Whenever tracing is not required keep the writing stylus away from the smoked drum.

(3) To verify the results, the experiment can be repeated with another fresh preparation of nerve muscle under the identical condition of previous experiment.

Experiment No. 2

Demonstration of the threshold stimulus, refractory period in a gastronemius nerve muscle preparation of frog

Theory: A threshold stimulus is the voltage of current or stimulus applied to the nerve, which is just sufficient to evoke a response. ***Refractory period*** is the duration, which a nerve once stimulated is unable to evoke any response on being stimulated again. This is actually the period during which a nerve recovers from the earlier stimulus. Just after the first stimulation, the second stimulus regardless of its strength is unable to evoke any response. This is called the ***absolute refractory period.*** This period is followed by the relative refractory period and during this period a second stimulus of much higher strength than the normal is capable of evoking response. When a relative refractory period is over, the nerve can be stimulated with the original strength of stimulus to evoke the response.

Procedure

(1) Prepare the gastronemius nerve muscle preparation from a freshly narcotized healthy frog as in case of exercises in muscular contraction. Transfer it in a Petri dish and keep it moist althrough the experiment.

(2) Set up the entire kymograph assembly as for muscle contraction; load the nerve muscle preparation in the muscle chamber.

(3) The speed of drum should be set at the lowest speed on speed panel of kymograph.

(4) Keep the short-circuiting key in closed condition.

(5) Make all electrical connection associated with kymograph, electrodes and student stimulator. Keep the switch of the student stimulator in off position but connect it with the mains.

(6) Load the nerve on the tips of the electrodes, placed very close to each other and both tips farther away from the muscle.

(7) Set the needle at the lowest voltage, on voltage dial of the front panel of student stimulator.

(8) Switch on the pulse switch.

(9) Start rotating the drum at a very slow speed from the seam.

(10) Put the switch of student stimulator and see whether there is any contraction in muscle or not. If no response that means the stimulus applied is much below the threshold value.

(11) Immediately change the voltage needle at next highest point and if you get a response in muscle then switch off the student stimulator, that means this is the threshold value of the current, which can evoke the response. Note it down.

(12) Start recording afresh with this stimulator strength and again push the pulse switch soon after the first response, if you do not get any response, then it means the nerve is in absolutely refractory period.

(13) Wait for sometime and then push the pulse switch and see the response, if the twitch formed this time is not equivalent to original one that means the nerve is in relative refractory period; in other words, it has not yet recovered from the first stimulus.

(14) Wait for some time more and then press the pulse button, you will notice that the twitch formed this time is approximately of the same magnitude or slightly bigger, thus indicating that the nerve has completely recovered, *i.e.* refractory period is over.

(15) Switch off the student stimulator, remove the writing stylus away from the drum and take out the drum from kymograph. Make a time scale with tuning fork.

(16) Remove the paper and label all tracings properly and fix the paper for permanent record.

(17) Note down your observation in a tabulated form and draw conclusion.

Experiment No. 3

Demonstration of the injury potential of a muscle using the gastronemius nerve muscle preparation

Theory: The mechanism of the creation of a propagated action potential in muscle cell, caused by a stimulus is well established and so is the fact of the existence of a difference of potential between the interior of a muscle and its outer surface meaning thereby that when the interior of muscle is connected with its outer surface, there is a flow of electrical potential.

Procedure

(1) Take a freshly narcotized healthy frog and dissect out its both gastronemius nerve preparation taking all precautions during dissection as in earlier cases. While dissecting out the preparation, try to take out the maximum length of the sciatic nerve attached to the muscle at neuro-muscular junction.

(2) Transfer both these preparations in a large shallow Petri dish and keep the preparations just moist. Extreme care is required in handling the preparations during summer months to avoid desiccation. Just mark the muscles as A and B.

(3) Place the two preparations in such a position so that the neuro-muscular junctions lie opposite to each other as shown in Figure 12.4 but take care that the sciatic nerve of both the preparation and muscle do not come in contact till desired.

(4) Make a V shaped cut through the middle of the muscle marked A with help of sharp razor blade.

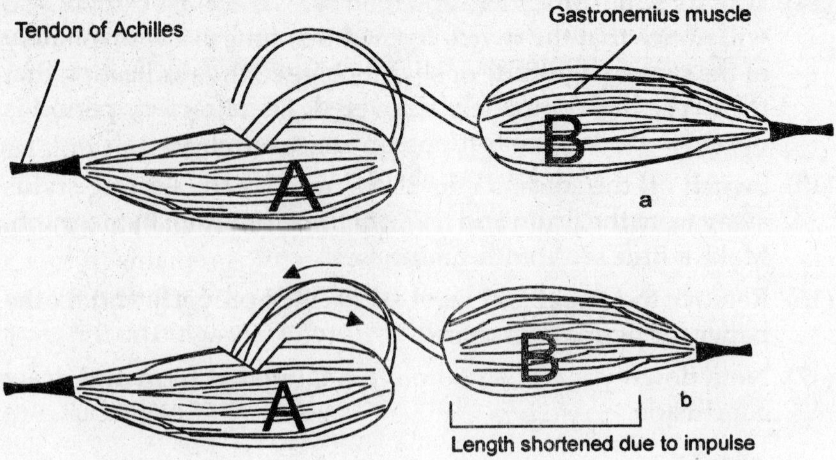

Fig. 12.4. Diagram to illustrate the passage and effect of injury potential

(5) Thereafter with the help of forceps gently lift the sciatic nerve of the uninjured muscle B and bring its free tip in contact with the injured muscle A as well as a part of it in contact with the nerve of injured muscle.

(6) Gently lift the nerve of the injured muscle A and bring it in contact with the cut surface of the same muscle. No sooner the nerve ending of the injured muscle comes in contact with the injured part, there is a sudden contraction of the uninjured muscle B as will be evident by the provisional shortening in its length.

Observations and Conclusions: The observations made above indicate that there exists a potential difference between the interior of a muscle and its external surface. It also indicates that a contact of the interior of muscle with external surface results in the propagation of an electrical potential.

Experiment No. 4

Demonstration of action current of nerve using gastronemius nerve muscle preparation of frog.

Theory: When the action potential of a muscle on contraction induces the contraction of another muscle through the nerve affected to it, it is referred to as the ***action current of nerve***.

Procedure

(1) Take a freshly narcotized healthy frog and make two

gastronemius muscle-nerve preparations and label them as A and B. This marking can be done on the Tendon of Achilles by means of copying pencil or just by sticking A and B by means of cello-tape.

(2) Transfer these preparations on dry and clean glass plates placed over a black paper sheet.

(3) The two preparations should be placed in such a fashion (Figure 12.5.) so that the sciatic nerve of A remains in touch with the muscle B at least at two points. The sciatic nerve of the muscle should lie free in the straight condition on the glass plate.

(4) The muscle-nerve preparation should be kept just moist, not very wet, with frog's Ringer's solution.

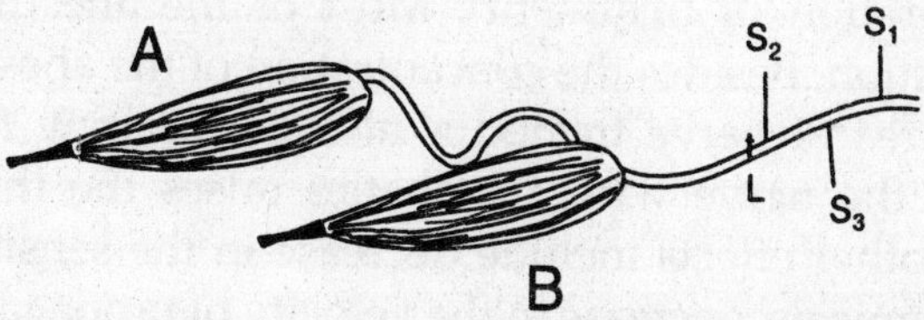

Fig. 12.5. Diagram illustrating the steps involved in demonstration of action current of nerve (S_1, First stimulation (a), S_2, Second stimulation (b), S_3, Third stimulation (c), L. Ligation

(5) When all the arrangements in the placement of nerve have been made then, first in attempt A the sciatic nerve of the muscle B should be stimulated at its distal free end with a median single induction stock or by using the microelectrodes of a student stimulator and observe that there is contraction in both the muscles A and B.

(6) In the second attempt (b), with the same preparation maintaining the original arrangement of nerve placement, stimulate the nerve of muscle B at its middle point by tetanising current, *i.e.* stimulation of the nerve at very rapid frequency. Here again you will observe the same effect as above in both the muscles.

(7) In the third attempt (c), with same placement of muscles and nerves, tie a ligature moistened with Ringer's solution at the point of tetanising stimulus in nerve of muscle B and stimulate the nerve towards distal end away from the ligature. You will see that there is no contraction of muscles.

Observations and Conclusions: Since, in third attempt (c) on putting ligature there is no contraction in both the muscle; therefore the contractions observed in (a) and (b) earlier to ligaturing could not be made due to the spreading of current of the induction coil or that caused by main current, but only due to the action potential of muscle, causing depolarisation in the sciatic nerve of muscle A, which was touching the muscle B at two places; thus causing the contraction of muscle.

Experiment No. 5

Demonstration of the affect of anesthesia on the conductivity of the nerve using gastronemius nerve muscle preparation of frog.

Theory: Anesthetics, such as cocaine, produce reversible effects while chloral hydrate produce less reversible effects (depending upon the duration of treatment). Most of the anesthetics block nerve conduction. Besides the concentration of the anesthetic compound, the area of nerve treated is also the limiting factors. The treatment of the nerve with anesthetics raises the threshold for stimulation, other effects include decrease in the sensitivity to respond for chemicals, decrease in the velocity of propagation and an increase in the refractory period. Summing up the whole it can be conclusively inferred that anesthetics hamper the excitability of the neuron.

Procedure

(1) First of all, take a piece of hard board or cork sheet measuring 4″ x 8″ or 4″ x 10″ depending upon the entire length of the nerve muscle preparation.

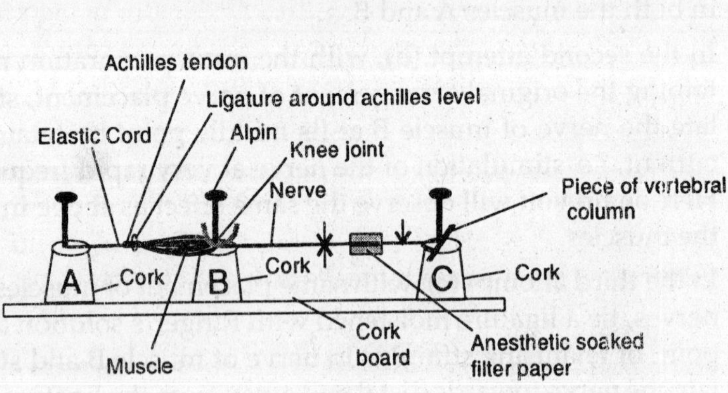

Fig. 12.6. Diagram to illustrate the effect of anesthetic on the conductivity of nerve

(2) With the help of some quick drying fixative, fix up on the cork board three corks with broader end downwards (as shown in Figure 12.6) at point A on one end, then fix the second one at a distance of 3″ at point B and thereafter the third one from B at a distance of 4″ or 5″ at point C. Ensure that all the corks are firmly fixed before using it.

(3) Make a gastronemius nerve preparation from a freshly narcotized healthy frog and then transfer it on a glass plate.

(4) Take a 3″ or 4″ piece of thin elastic cord and make a tight ligature around the tendon of Achilles as far away from the muscle, as possible.

(5) Lift the preparation and fix up the knee joint with the help of alpin at the top of cork B taking care neither to injure the nerve nor to stretch it.

(6) Make a loop at the free end of elastic cord so that in the fully extended condition of the muscle it lies in the centre of cork A and place this loop in the alpin already fixed in the centre of a cork A. In this condition, the elastic cord and muscle should be in fully extended condition almost parallel to the surface.

(7) Pick up the piece of vertebral column, extend the nerve upto the cork C and fix it up with the help of one or two alpins. While fixing it, the stretching of nerve should be avoided.

(8) Always keep the muscle and nerve moist with Frog's Ringer solution.

(9) Stimulate the nerve nearer to vertebral column with the help of a pair of forceps with a medium strength single induction coil or by the electrodes of a student stimulator using a single pulse and that to once only. If you observe the contraction of muscle that means the loading of both muscle and nerve has been perfect.

(10) Take a strip of filter paper, approximately 0.5cm in width, moist it with Ringer's solution, roll round the nerve nearer to the cork C. This rolled strip should be kept in position by putting a tight ligature by means of sewing thread.

(11) Again stimulate as in step no. 9. If contraction of muscle takes place that means this placement of strip is also perfect.

(12) Moisten the filter paper strip either with one or two drops of 0.3 to 0.5% solution of Novocain or cocaine and wait for 20-30seconds. Apply a single shock stimulus on the nerve seg-

ment between the filter paper strip and the cork C. If no contraction of muscle takes place that means the passage of current has been blocked on account of local anesthesia of the nerve at the point of application of anesthetic.

(13) Stimulate the nerve segment in between the filter paper strip and cork B. This stimulus should be away from the filter paper strip. The contraction of the muscle will confirm that the passage has been definitely blocked by the local anesthetization of the nerve segment due to cocaine or Novocain anesthetics.

Precautions

(1) If after applying the 0.3 or 0.5% solution of cocaine or Novocain and waiting for 20-30 seconds on stimulation, the contraction of muscle is noticeable then first of all you should wait for a few more seconds and then again stimulate. If no contraction occurs then proceed ahead. If contraction takes place then increase the strength of anesthetic solution.

(2) The nerve should always be kept moist with Ringer's solution.

(3) If the length of nerve is sufficient enough then the experiment can be repeated by applying the anesthetic on the nerve segment at some point between the first filter paper strip and the cork B.

Experiment No. 6

Determination of reflex activity in human beings.

REFLEX ACTIONS: These are the fixed patterns of response to a particular stimulus and they need not involve an awareness of stimulus. The familiar knee jerk action is a two-neuron reflex; the afferent neuron synapses directly with the efferent neuron without the involvement of and no internuncial neurons.

There are many actions, which are instantaneous and are not under control of will or consciousness. A common example is the instantaneous withdrawal of the hand when it accidentally touches a hot stove. In this situation we do not make a judgement to withdraw the hand. Any such action or response is called ***reflex***.

The reflexes are broadly classified into three categories, namely: (1) **Natural reflex**, in this type no previous experience or learning is required. Such reflexes are inheritable. (2) **Acquired reflex** is one, which is developed in ones life due to experience or learning. For example, when you simply see or smell the tasty food, you experience watering (salivation) of your mouth. (3) **Conditioned reflex,** is exemplified by the onset of salivation, when you start to chew or eat food. This is different from the salivation under natural reflex because in the earlier example the sight or smell of food was enough for response.

Some more Common Reflexes in Humans beings: These are either the simple reflexes or the conditioned reflexes, but all these are associated with spinal cord.

The Simple Reflexes include

(1) Knee-jerk, in which the leg is involuntarily extended as a result of sharp tap below the kneecap in a relaxed leg rested on the other leg while sitting.

(2) Sudden closure of eyelids, when an object suddenly approaches the eye or when a strong beam of light is flashed across.

(3) Withdrawal of hand or any part of the body when suddenly pinched.

The Conditioned Reflexes include most of our daily habits and acts, which fall under this category, *e.g.*

(1) The tying of your shoelaces while talking and not knowing whether you are first putting the right lace or left or vice-versa.

(2) Playing a musical instrument.

(3) Using keys of the machine while tapping.

(4) Giving a hand signal to your right automatically when turning your cycle or car to the right.

(5) When you ride a bicycle or drive a motorcar, you suddenly find an obstacle in front or the green light of the traffic turns red, you immediately apply brakes without thinking.

Physiology of Reflex action: Since a reflex action occurs in twinkling of eyes, hence the pathway for receiving and sending information must be short. In a simple reflex, impulses are not conducted up and down the spinal cord but leave at the same level at which they enter (Figure 12.7). This may involve two neurons or three neurons. When it involves two neurons, these are a dorsal (afferent) neuron located on the dorsal root and the ventral motor (efferent) located in the ventral part of the grey matter of the spinal cord with its fibre in the ventral root.

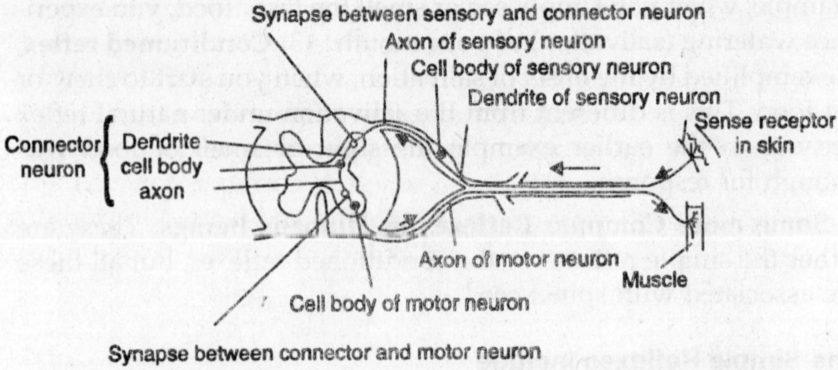

Fig. 12.7. Diagram of a reflex arc. showing the pathway of an impulse, indicated by arrows.

Sometimes a third neuron (called intermediate or internuncial neuron) is also involved in the reflex arc, lying between the sensory fibre and the motor neuron inside the spinal cord.

Summing up the whole, it can be concluded that the spinal reflex is the simplest, quickest pathway by which action can be taken in response to a stimulus. The actions include either the movement of entire body or a part of its body away from the threatening stimuli, such as heat from the stove or prick of a sharp instrument, electrical shock or irritation from the chemicals, *etc.* The effects of different kinds of stimuli can be demonstrated by performing experiments on a spinal cord of frog.

Experiment No. 7

Demonstration of the development of a reflex action in response to different types of stimuli in frog.

Procedure for the preparation of spinal frog

Take a healthy frog and wash it with water and then hold it with a piece of rough cloth through its belly. Place it in the left hand with its belly downwards. Slide its head between the first and second fingers and flex the head sharply downwards in such way that the head makes a right angle against the trunk. Prick a metallic pithing needle through mid-longitudinal line of the frog in front backward until you feel the presence of a depression, which will be, situated approximately 3-4 cm (depending upon the size of the frog) behind a transverse line adjoining the posterior borders of the tympanic membranes. Now, with sufficient pressure insert the tip of the needle in the foramen magnum and push it towards the anterior part of the skull. Thereafter, start twisting and turning the needle from side to side and rotate it once or twice with a view to destroy as well as to disconnect the spinal cord with medulla oblongata. This treatment will render the frog non-responsive to pain as such. The complete and successful pithing becomes evident when the frog fails to close the eyes, even when its cornea is touched with something. In this way, a frog is prepared and is used in studying various types of simple and complex spinal reflexes due to the effect of different types of stimuli.

Experimental Procedure

For performing the experiments on spinal reflexes, a S-shaped metallic hook is tied in the lower jaw of the frog and then hang the frog through a non-elastic thread from the clamp mounted on a stand and proceed as under to study various types of spinal reflexes under the influence of different types of stimuli.

Take a sharp forceps and gently pierce or pinch any one of the toes of the left leg. You will see that the leg is lifted a little above. Now, pinch harder with pressure. This time you will see that the left leg is lifted much above from the original position. This will indicate the action of the stimulus as well as the action of the increasing strength of the stimulus. The response obtained shows the reflex action. Similarly the reflex actions can also be studied using different types of stimuli of varying nature, such as electrical, thermal, chemical, *etc.*

13

Physiology of Excretion

INTRODUCTION

The main organs of the body concerned with the removal of waste products of metabolism are the kidneys and the lungs. The main functions of the kidneys are to eliminate water, salts and the waste products of metabolism and to regulate the pH of the blood and tissue. Most of the nitrogenous waste products are eliminated through the kidneys. An adult on an average excretes in the urine about 15grams of nitrogen per day. The volume of urine excreted in 24hours by a normal individual on an average is about 1200ml. There are many drugs that have diuretic effect, *i.e.* they stimulate kidneys and cause an increased production of urine. Caffeine, a substance found in coffee and tea, has similar effects.

General Characteristics of Urine

Appearance: Normal urine has a yellow colour ranging from straw to amber. It is usually clear and transparent when voided. On standing most of the samples develop a cloud composed of nucleo-proteins or mucoid. All these are discernible by simple vision.

Odour and Taste: Urine has a faint aromatic odour. Normally urine has a salty taste because of the presence of sodium chloride. However, both the odour and the taste vary with different foods, drugs, and diseased conditions. The urine of a diabetic patient tastes sweet because of the presence of glucose. Before the days of the clinical laboratories, tasting the urine for sweetness was the practice in diagnosis of diabetes. The reaction of urine varies over a wide range, depending on several factors. As a rule, it is acid, having a pH about 6.0. Since, it also shows variation with various physiological diseases so its qualitative and quantitative measurements also play a significant role in the diagnosis of certain diseases.

Specific gravity: The specific gravity of urine normally ranges from 1.015 to 1.030 under ordinary conditions. The specific gravity of 24hour sample is 1.020 (1.016-1.025). These values may vary considerably, depending on the water intake. As a rule, the specific gravity is inversely proportional to the volume of a 24hours sample. The specific gravity of a urine sample also plays an important role in diagnosis of certain diseases, so its measurements in pathological laboratories are commonly carried out as under:

Exercise No. 1

Determination of the specific gravity of a urine sample

Apparatus and accessories: Hydrometer, measuring cylinder, litmus paper and thermometer.

Theory: Specific gravity is measured by means of a special type of hydrometer known as ***urinometer*** (Figure 13.1). It is a glass instrument consisting of a graduated stem weighted with a mercury bulb, used for measuring the specific gravity of urine. Since it is weighted with mercury, it floats in the urine so that the calibration, which corresponds to the surface level of urine, is red. The lower the specific gravity, the further the urinometer sinks. Since the instrument is usually calibrated at 15°C, it is necessary to correct for temperature. This is done by adding 0.001 for every 3°C above 13°C in reference to room temperature and subtracting 0.001 for every 3°C under 15°C in reference to room temperature.

For example, if the urinometer at 21°C measures the specific gravity at 1.018 then according to principle of calculation at 15°C the specific gravity will be equal to 1.018 + 0.002 = 1.020. The normal values of specific gravity as per urinometer calibrated at 15°C vary from 1.015-1.025.

Procedure

Take a measuring cylinder and pour the urine sample to approximately three quarters of the capacity or more of the cylinder. Allow it to stand as such till its temperature equalizes to room temperature, which has been ascertained by a thermometer. Now put the urinometer in the urine sample contained in the measuring cylinder taking care that its sides should not touch the sides of the cylinder. Note down the specific gravity by reading the graduated markings coinciding with the upper level of urine and have a record of it.

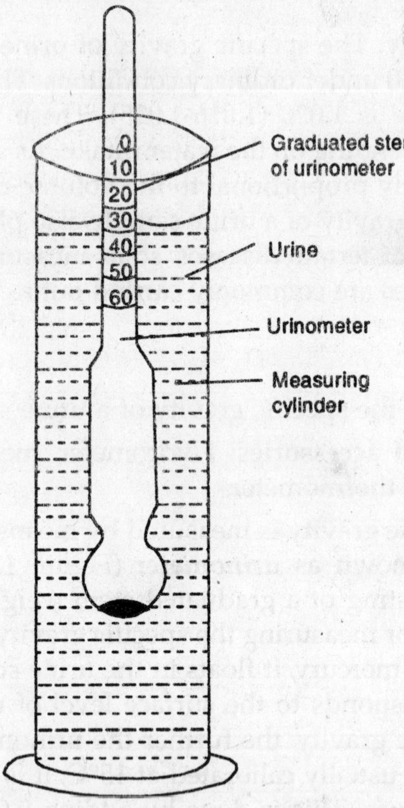

Fig. 13.1. Urinometer

Precautions

(1) The urinometer should not touch the sides of cylinder.

(2) Zero of new instrument should always be checked before use.

Exercise No. 2

Measurement of specific gravity of a small amount of urine by weighing method.

Take the sample of urine in a specific gravity bottle of suitable size or use of pipette to know the exact volume of urine sample and find out the weight of urine. Now find out the weight of water equal to the volume of urine sample. The ratio of these two will be equal to the specific gravity of urine.

Exercise No. 3

Determination of pH of urine.

Apparatus and accessories: Litmus paper, small watch glasses or cavity blocks with lids, sample of urine of infants, children or adults (normal males and females) with different food habits, of diseased persons like those suffering from diabetes.

Procedure

Take each sample in a small cavity block and put a label on it. Using a separate litmus paper strip, test the sample simply by dipping litmus paper in the sample and matching the colour developed on it, with that shown on colour index strip, find out the pH and record it in the table on the format given below:

Table 13.1: To record pH values determined by the student.

S. No	Detail of sample	pH	Acidic	Basic	Conclusion	
					Increase	Decrease
1.	Infant urine					
2.	Male adult – Veg. diet					
3.	Male adult – Nonveg. diet					
4.	Female adult – Veg. diet					
5.	Female adult – Nonveg. diet					
6.	Female adult – during pregnancy					
7.	Male adult – diabetes mellitus					
8.	Female adult – diabetes mellitus					
9.	Male adult – diabetes insipidus					
10.	Female adult – diabetes insipidus					

It is not necessary to have all samples at one time. Samples can be collected from different places as per convenience and be tested from time to time in the laboratory.

Exercise No. 4

Determination of the amount of solids in a given urine sample.

Apparatus and accessories: As in exercise no. 1.

Theory: An idea of the amount of solid also helps a physician in confirming certain disease as, for instance in case of diabetic patient.

Table 13.2: Composition of a 24-Hour sample of urine of a normal person.

Name of constituent	Grams
Water	1200
Total Solids	60
Urea	30
Ammonia	0.7
Uric acid	0.7
Creatinine	1.2
Hippuric acid	0.7
Amino acids	0.2
Phenols	0.2
Chlorides, like NaCl	12.0
Sodium	4.0
Potassium	2.0
Calcium	0.2
Magnesium	0.15
Total sulphates as SO_3	2.5
Inorganic sulphates as SO_3	2.0
Ethereal sulphates as SO_3	0.2
Unoxidised sulphur as SO_3	0.3
Phosphates as P_2O_5	2.5

Procedure

Find out the specific gravity of the sample of urine as per procedure followed in previous exercise. To determine the total solids in a litre of urine sample, multiply the second and third decimal figures of the specific gravity of the urine sample measured above by 2.6. It was Long who suggested this method so value of 2.6 is known as Long's coefficient. For example, a person whose 24hour urine sample has the specific gravity of 1.025, shall have total solids amounting to 25 x 2.6 = 65.0 grams.

Exercise No. 5

Detection of urea using Hypobromide method.

Theory: When urea is treated with hypobromide, a certain volume of nitrogen depending upon its amount is liberated, which is measured by Gerrard urinometer (Figure 13.2).

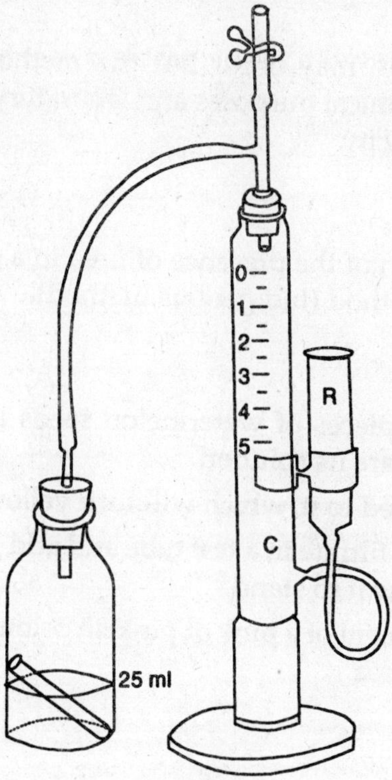

Fig. 13.2. Gerrard – Urinometer

Procedure

Take a wide mouthed glass bottle bearing calibrated markings. Pour 25ml of alkaline hypobromide solution in this bottle. Add just a few drops; say 4-5 of caprylic acid to minimize frothing to this. Take a small test tube and pour about 5ml of urine sample in it and carefully place this test tube inside the bottle in a slanting position as shown in Figure-13.2. To avoid the risk of breaking of the test tube, while placing inside the bottle, fix a small piece of rubber tubing on to the bottom end of the test tube. Now open clip A and allow the water in the cylinder C to rise upto the zero by raising the height of the reservoir R and close the clip. Thereafter, lower the reservoir a little low, the water level will fall to some lower height but will remain stationary ensuring no leakage. Now gently tilt the bottle so that the mixing of urine with hypobromide solution takes place. Shake it carefully when evolution of nitrogen is topped, level the bottle again and read of the percentage of urea on the urinometer, which is calibrated so as to read directly in term of grams urea per 100ml of urine.

The error upto 5% may occur but the method is sufficiently accurate for many clinical purposes and laboratory experiments on account of its simplicity.

Exercise No. 6

Qualitative detection of the presence of urea in a given urine sample using urease method (indigenous method).

Test No. 1

(1) Grind a few pieces of watermelon seeds and add distilled water to prepare its solution.

(2) Add phenol red to it, which will turn yellowish and filter it.

(3) Take 2-3ml of filtrate in a test tube and add 1ml of urine sample to it. Allow it to stand.

(4) The development of a pink or pinkish colour will indicate the presence of urea.

Test No. 2

(1) Place the knife point of Jack bean meal in a test tube.

(2) Add 4-5 drops of phenol red and 2ml of distilled water to it.

(3) Mix by agitation and allow the mixture to stand for 10-15minutes and then filter it.

(4) In a test tube take 2-3ml of urea and add some of the filtrate (enzyme solution) to it.

(5) Mix the two thoroughly well and allow to stand at room temperature and observe that the mixture is rendered red in colour due to the conversion of urea into ammonia.

$$CO(NH_2)_2 + H_2O \rightarrow CO_2 + 2NH_3$$

It is important to adjust the initial pH 7, when testing for small amount of urea.

Test No. 3

The urea is an important nitrogenous constituent of urine and is best determined by converting into ammonia by means of the enzyme urease, which is commercially available. The amount of urea in a 24hour sample is about 30 grams.

In a test tube, take about 3-5ml of urine sample. Add about 3-5 drops of phenol red to this. Wait for the development of pink colour, failing which add 1% NaOH solution drop by drop until the solution is alkaline. This should be ascertained by a litmus paper. Thereafter, add a little quantity of 1% acetic acid and wait for the disappearance of pink colour developed above. At this stage the pH of the mixture in test tube should be above 7.0. Now, the temperature of the mixture should be gradually raised to about 40°C by heating the test tube over the flame of a burner or spirit lamp. Take a urease tablet and crush it to form its powder. Add half of this powder to the mixture in test tube; slowly shake it at short intervals. By this treatment the mixture will become alkaline, its colour will change from yellow to red and after a short while to reddish purple, for the reason that neutral urea (pH 7.0) is converted into alkaline ammonium carbonate.

Precaution

Since pH plays an important role, therefore, extra care has to be taken when adding acetic acid.

Exercise No. 7

Detection of urea in urine sample by xanthydrol method

Procedure

Take 2ml of fresh urine sample in a clear test tube and add 6-7ml of glacial acetic acid to it. Then also add 2-3ml of 10% solution of xanthydrol (prepared in methanol). Allow it to stand for sometime and watch the development of copious amount of crystalline precipitate. The precipitate of xanthyl-urea formed in this way indicates the presence of urea in urine sample under test.

Exercise No. 8

Detection and determination of uric acid in urine.

Theory: Uric acid is one of the important nitrogenous constituents of urine. It is usually present in the urine as urates.

Test No. 1

Take 1-2ml of urine in a test tube and acidify it with a few drops of concentrated HNO_3. The uric acid will be set free from the urates present in urine. Allow it to stand for sometime, a sediment of red crystals will indicate the presence of uric acid.

Test No. 2

If the uric acid has already crystallized in the sample of urine, take a little bit of this solid material in a small porcelain dish or crucible, add 4-5drops of concentrated HNO_3 to it and then slowly heat it over the low flame of the burner until it is dry. Now, remove it from the flame and add diluted solution of liquor ammonia. Wait for sometime. The development of reddish violet colour in the residue will indicate presence of uric acid.

If on adding a few drops of NaOH or KOH, reddish violet colour turns into blue-violet, the presence of uric acid is confirmed.

Exercise No. 9

Detection of creatinine in the urine.

Theory: It is present in the form of an anhydride of creatine and it ranges from 1-1.25gm in a 24hours sample of urine of an average adult. It is higher for men than for women and children. Its intake in food usually increases its amount in urine. It may be noted that the determination of creatinine in the diagnosis of diseases is of insignificant value.

Test No. 1 Sodium nitropruside Test

Theory: This test is based on the appearance of a ruby red colour when creatinine is treated with an alkaline solution of sodium nitropruside.

Procedure

Take 10ml in a test tube and place in it 5ml of urine sample. Add the freshly prepared 0.1M sodium nitropruside solution (prepared by dissolving 15g of sodium nitropruside in 500ml of distilled water) to it drop by drop. Now add M sodium hydroxide solution to it drop by drop till it becomes alkaline, which has to be ascertained with the help of a litmus paper. Gently shake and allow to stand. A ruby red colour will soon develop indicating the presence of creatinine, which may subsequently turn to yellow with further lapse of time or on heating.

Test No. 2 Creatinine Picric-acid Test

Theory: Creatinine forms creatinine picrate when mixed with picric acid, which on becoming alkaline leads to the development of a deep yellow or red tautomeric compound.

Procedure

Take 4ml of urine sample in one test tube, marked A and an equal volume of distilled water in another test tube marked B. Now with the help of a pipette add 1ml of saturated solution of picric acid in distilled water as well as 1ml of 2M sodium hydroxide in each test tube separately. Allow both the test tubes to stand and notice the development of different colour in test tube A, indicating the presence of creatinine.

Exercise No. 10

Detection of uric acid in animal tissue extract by using Benedict's method.

Procedure

(1) Dissect the cockroach and take out the fat bodies or malpighian tubules in distilled water.

(2) Grind the tissue in a grinder until suspension.

(3) Take 3-4ml of the suspension in a test tube and add 1ml of Benedict's reagent and some saturated solution of Na_2CO_3 to it.

(4) The development of a deep blue colour, produced due to the reduction of the reagent by uric acid in alkaline solution, will indicate the presence of uric acid.

Exercise No. 11

Histochemical detection of uric acid crystals by using $AgNO_3$ formalin method.

Solutions required

(A) Prepare 4.4% solution of formalin and make it neutral by adding a few pieces of chalk stick until effervescence appears. When bubbling is subsided that means solution is ready for use.

(B) Prepare 1% solution of $AgNO_3$.

Procedure

(1) Take out the fat bodies and malpighian tubules of cockroach and put them in a cavity block.

(2) Immerse the material in a mixture of equal volumes solutions A and B. If the materials keep floating, submerge them with the forceps.

(3) Cover the cavity block by black paper and keep it in dark for 6hours.

(4) The uric acid crystals will become black.

(5) Treat the material with distilled water, dehydrate clear and mount.

(6) The presence of black coloured uric acid crystals under microscope will indicate presence of creatinine.

The detection of the pathological constituents of urine including sugar, fructose, pentose, albumin, acetone, *etc.* is carried out by specific tests as detailed below.

Exercise No. 12

Detection of glucose in urine sample by Benedict's Test

Preparation of Benedict's reagent

(i) Dissolve by heating about 173g of sodium citrate and 100gm of anhydrous sodium carbonate in approx. 600ml of distilled water in a large beaker.

(ii) Separately prepare a solution of copper sulphate ($CuSO_4 \cdot 5H_2O$) by dissolving 17.3gm of copper sulphate in 100ml of distilled water.

Slowly mix the two solutions by adding and continuous stirring. Cool the mixture and transfer it from beaker to a volumetric flask of 1litre capacity and make to the mark with addition of more of distilled water.

Test No. 1 Benedict's Test for glucose

This is a very sensitive test and will yield positive results even if the amount of sugar present is very small, say for example 0.2% solution of glucose in distilled water.

Take 8-10 drops (or 0.5ml) of urine sample in a test tube, add 5ml of the Benedict's reagent to it and boil for 2minutes or place in a boiling water bath for 3minutes. Allow it to stand and cool down spontaneously. A precipitate will be formed, whose colour will vary from greenish to yellowish brown to reddish brown, depending upon the amount of reducing sugars present. A white precipitate is developed if significant quantities of phosphate are present. Satisfactorily positive results are obtained if glucose is present.

Test No. 2 It is a very simple method of Benedict's test for sugar in urine.

Take 5ml of Benedict's solution in a test tube and add about 40drops of urine sample to it. Boil for 2minutes and allow the test tube to stand and cool down to room temperature. The development of a brick red or deep yellow precipitate indicates the presence of glucose.

Exercise No. 13

Detection of glucose in urine using Fehling's Test.

Preparation of Fehling's reagent

Method I

Solution A. Copper sulphate solution: Dissolve 17.30gm of $CuSO_4$ in a small quantity of distilled water and make upto 250ml.

Solution B. Alkaline Tartarate solution: Take 63gm of KOH and 86gm of Rochelle salt and dissolve in small quantity of distilled water. Then bring the volume to 250ml by addition of distilled water. For preparing Fehling's reagent, mix equal volume of Solution A and B.

Method II

Solution A: Dissolve 6.98gm of $CuSO_4.5H_2O$ (copper sulphate) in 100ml of distilled water.

Solution B: Dissolve 25gm of potassium hydroxide and 34.6gm of Rochelle salt (sodium potassium tartarate) in 100ml of distilled water.

Mix equal volumes of solutions A and B when required. This is done to prevent deterioration. Before using the Fehling's solution, verify its validity. In a test tube take equal volumes of solutions A and B and dilute with two volumes of distilled water and boil for a second. If the solution remains clear it means no deterioration, else it should be discarded.

Test No. 1

Theory: This test is based on the reduction of copper salts to cuprous oxide by sugars, which contain a free aldehyde or ketone group.

Procedure

Take 5ml of urine sample in a test tube and add 1ml of the Fehling's solution to it. Gently shake it and boil for a few minutes. This will result in the development of a characteristic brick red precipitate of Cu_2O. In case the precipitate formed is of yellow colour due to the temporary formation of copper hydroxide, further boiling will impart a brick red colour indicating the presence of reducing sugars.

Test No. 2

Separately take 4ml of Fehling's solution and 4ml of urine sample in two test tubes. Bring both to the boiling point. If no precipitate is formed in Fehling's solution then add urine sample of the other test tube and observe. Transfer equal volumes of this mixture in two separate test tubes and once again bring them to boiling point. The amount and colour of the precipitate formed will give a rough idea of the amount of reducing substances present. A greenish yellow colour will indicate presence of small amounts; a yellow deep reddish brown to larger amounts.

Exercise No. 14

Detection of glucose in urine using Clini test.

The alkaline copper reagent is available in tablet form. Each tablet contains copper sulphate, caustic soda (38.5%), sodium bicarbonate and citric acid.

Take 5drops of urine sample in a test tube and 10drops of distilled water and one tablet alkaline copper reagent. Compare the colour developed with the colour indicated on colour chart and find out the approximate amount of sugar present in the test sample.

Exercise No. 15

Detection of glucose in urine using Clinistix test

Theory: This is a specific test for glucose. The strips carrying reagents are commercially available with medical stores and chemists. Half of each strip is impregnated with glucose oxidase, o-toluidine and a vegetable peroxidase.

Procedure

Take a strip and dip the test end of the strip in a test sample of urine contained in a small Petri dish and remove it immediately. The immediate development of a blue colour within a minute or so gives a positive indication of presence of glucose. With this technique positive results can be achieved with concentrations of glucose above 10-100mg/100ml. This is a valuable technique for screening to exclude the presence of significant amount of glucose as for instance in out door patient and anti-natal clinics but on account of its great sensitivity, the final confirmation should always be made with Benedict's Test or Clini Test.

Exercise No. 15

Detection of fructose in a urine sample using Seliwanoff's Test for Fructose

Preparation of Seliwanoff's reagent

Take 16.5ml of concentrated hydrochloric acid in a beaker and dissolve 25mg of resorcinol in it. Transfer the mixture in a measuring cylinder and by adding distilled water bring the volume to 50ml.

Theory: Treatment of fructose with hydrochloric acid results in the formation of derivatives, called furfuraldehyde, which on being linked with resorcinol produces a red coloured substance.

Procedure

Take 5ml of Seliwanoff's reagent in a test tube and add about 0.5ml of filtered urine sample to it. Heat it to boiling point. If a red colour develop within 30seconds that means fructose is present. In the absence of glucose, the sensitivity of test lies in the range of 0.1 percent; but in presence of large amounts of glucose the test is less sensitive.

Exercise No. 16

Detection of pentoses in urine sample using Bial's Test

Preparation of Bial's reagent

Take 100ml of concentrated hydrochloric acid in a measuring cylinder, dissolve 800mg of orcinol in it and add 5drops of 10% solution of ferric chloride.

Theory: Hydrochloric acid acts on pentoses in more or less the same fashion as on fructose but the substances yielded react with orcinol to produce a green coloured compound.

Take 0.5ml of filtered urine sample in a test tube and mix it with 5ml of Bial's reagent. Heat the test tube until the mixture starts boiling. The development of a green colour will indicate the presence of pentoses. The test is sensitive to 0.1% solution of pentoses. The development of red colour in this case will indicate the presence of fructose in large amount.

Exercise No. 18

Detection of reducing substances in a given sample of urine by paper chromatography.

Apparatus, accessories and reagents: Chromatographic chamber, Whatman No. 1 filter paper.

Solvent system: Isopropanol: water; 160: 40 (v/v). This is most pleasant to use and is not affected much by variation in temperature.

Staining reagent: Aniline-diphenylamine. (Mix 10 volumes of 1% v/v solution of aniline and 1% (w/v) diphenylamine in acetone with 1volume of 85% phosphoric acid.

Standard samples: A composite standard containing xylose, fructose, glucose, galactose and lactose for routine use with additional sugars, if required, should be prepared.

Theory: Due to low R_f values of sugars, the descending type of paper chromatography is done. For this a tank preferably 20" in height is most suitable but there is no problem in using ascending type also.

Procedure

It is more or less the same as described in chapter 5 on paper chromatography, however with a view to avoid any confusion, the main guidelines are given below.

Cut a strip of Whatman No. 1 filter paper according to dimensions of tank preferably 50cm in length. Draw a base line and place spots of standards at significant distances from each other. Now place a spot of urine sample (approximately 5 and 50μl), depending upon the amount of reducing substances present. A rough idea of the same is obtained by Benedict's test. Apply the spot of standard in the range of 5 or 10µl. Now, place the strip in solvent system to run for 16-17hours at room temperature.

When the solvent front has travelled a significant distance, remove the strip from tank and blow off the solvent in a stream of air or by means of a hair dryer.

For developing the spots, either dip the strip in a bath of staining agent or spray it with staining reagent, such as aniline-phenylamine. Thereafter, heat it just for a few minutes at 95-100°C. The spots developed thus will reveal significant differences in colour, e.g. spots of pentoses are of grey brown colour, fructose, sucrose and galactose of greenish blue colour, lactose and maltose of blue colour.

Determine the Rf value of each spot and by comparing their Rf with the known Rf values as given in the table below, work out the nature of reducing substance for the solvent system used herein.

Table-13.3 Rf values in solvent system (Isopropyl: water, (v/v).

S. No.	Name of reducing substance	Rf value (Solvent system - isopropyl water)
1.	Xylose	1.55
2.	Ribose	1.39
3.	Xylose	1.30
4.	Arabinose	1.10
5.	Fructose	1.05
6.	Glucose	1.00
7.	Galactose	0.80
8.	Sucrose	0.60
9.	Maltose	0.47
10.	Lactose	0.32

Exercise No. 19

Detection of reducing substances in a urine sample by Thin layer chromatography (TLC).

Apparatus and accessories: Mainly as detailed in chapter 6 but for ready references as under:

(1) **Silica gel G** (E. Merck) on glass plates.

(2) **Solvent mixture:** n-butanol: acetic acid: water; 75: 25:6 v/v/v

(3) **Colouring agent:** aniline phenylamine, as in previous experiment.

(4) **Known standards:** Prepare by mixing equal volumes of 1% solutions of each, such as glucose, galactose, lactose in isopropanol. Roughly 1.5μl of this solution will contain 5μg of each.

Theory: Same as in chapter-6 on TLC. The results with paper chromatography are quite satisfactory but TLC is less time consuming and more sensitive, hence preferable.

The urine sample under test should contain the reducing substances between 2.5 and 7.5μg per 1.5μl-2μl. If values in urine are greater, then it could be diluted accordingly. Take a prepared chromatoplate of 10 x 20cms and on a base line 2cm away from the lower edge, apply spots of the standard, 1cm away on either of the spot of urine sample. At a time 2 or 3 urine samples can be applied with proper marking. Run the plate in solvent system in a glass jar already saturated with solvent system and covered by a lid to make it air tight for 2-3hours so that the solvent front travels to a distance approximately 15cm from the base line. When the run is finished, remove the chromatoplate and dry in hot air. For this, a hair dryer can be used. Spray with colouring reagent and dry the plates in a hot air oven set at 110°C for 5-6minutes.

Table 13.4: Rf values of known samples in a solvent system of n-butanol: acetic acid: water; 75:25:6 (v/v).

S. No.	Name of reducing substance	Known Rf value	Calculated Rf value	Result
1.	Lactose	0.14		
2.	Galactose	0.35		
3.	Fructose	0.39		
4.	Glucose	0.42		
5.	Pentoses	0.5-0.9		

Observations: Identify the reducing substances on the basis of colour reactions, as for instance, glucose, galactose and lactose produce grey colour spots, fructose a pink spot. Find out the Rf values of each and for confirmation compare with Rf values given in table 13.4.

Exercise No. 20

Preliminary test for the presence of proteins in the urine.

Take about 5-6ml of urine sample in a test tube and make it

slightly acidic by adding dilute acetic acid drop by drop. Test it with litmus paper. If urine is not clear then filter it.

Take another test tube and pour three quarters full of urine in it and by slightly tilting the test tube, heat the top one or two centimetres to boiling point. If proteins are present then coagulation will take place and it should be compared with the unboiled urine lying in the lower portion of the test tube. The distinction between the coagulated and uncoagulated urine is readily worked out if comparison between the two is made against a dark background.

If clearly satisfactory results with above tests for albumin are not obtained then it is necessary to isolate the protein from the urine sample before performing the test. For this mix equal volume of urine and saturated ammonium sulphate solution. Shake it well and allow standing. The albumin present remains in the solution whereas the globulins are precipitated and can be separated simply by filtering the solution.

Exercise No. 21

To detect the presence of albumin in the sample of urine.

Theory: Albumin is usually the main protein present in urine although appreciable amounts of globulins may also occur. The disorder is referred to as proteinuria or albuminuria.

Many tests are available for detecting the presence of proteins, such as, Heat Coagulation Test, Million's Test, Xanthoproteic test, Biuret test and Ninyhydrin Test (Triketo-hydrindene hydrate secretion). For immediate practice, a selected few are described below:

Test No. 1 Heat Coagulation Test for Albumin

It is most simple and common test. Take 4ml of urine sample in a test tube and heat the upper layer of urine over a free flame to boiling point. The development of clouding or the formation of the precipitate of earthy colour is an indication of the presence of albumin. Now, remove the test tube from the flame and add diluted acetic acid to it. This will result in the dissolution of the earthy coloured precipitate of phosphate formed above, whereas albumin will continue to cloud the sample.

Test No. 2 Million's Test for Albumin

Take 5ml of urine sample (or 5ml of dilute solution of egg albumin) in a test tube and add 4-5 drops of Million's reagent to it. Gently shake the test tube for uniform mixing. Now, gradually heat the mixture over a low flame of burner and bring it to boiling point. Initially white precipitate will be formed, which will gradually turn to red. In case no colour is developed, then a few more drops of reagent be added and process of heating continued further.

Note: Excessive use of reagent should be avoided from the point of view of accuracy of test.

Test No. 3 Xanthoproteic Test for Albumin

Method-I: Take 2-3ml of urine sample (or 5ml of dilute solution of egg albumin) in a test tube. Slightly tilt the test tube and add 2-3ml of concentrated HNO_3 from the side of the test tube. Observe the formation of white ring, which indicates the presence of albumin.

Method-II: Take 3-4ml of urine sample (or 5ml of dilute solution of egg albumin) in a test tube and add 1ml of concentrated HNO_3 (fuming nitric acid) to it. Initially a white precipitate will be formed on heating. On further heating this will turn yellow and finally it will undergo complete dissolution rendering the solution to appear yellow. Allow the test tube to stand and cool down to room temperature. Now, carefully add excess of ammonium hydroxide or sodium hydroxide. This will turn the yellow colour of solution to an orange red colour.

Test No. 4 Biuret Test for Albumin

Take 3-4ml of urine sample (or 5ml of dilute solution of egg albumin) in a test tube. Add an equal volume of freshly prepared 10% solution of sodium hydroxide to this. Mix it thoroughly well and then add 0.5% solution of copper sulphate ($CuSO_4$). The copper sulphate solution should be added very carefully, by adding drop by drop with simultaneous mixing, until a purple violet or pink violet colour is developed. The type and intensity of colour developed depends upon the nature of protein present in the sample.

Exercise No. 22

To detect the presence of acetone bodies in the sample of urine using Sodium Nitropruside Test or Rother's Test.

Theory: Normal urine contains traces of acetone bodies, the quantity of which varies from 20-30mg of β-hydroxybutyric acid constituting the greatest part of the total acetone bodies. Acetone is usually tested in a routine examination of urine in pathological laboratories.

Rother's Test

Method-I: Take about 2-3ml of urine sample in a test tube and add an excess of solid $(NH_4)_2SO_4$ to it so that the urine sample becomes saturated with it. Now, pour 2-3drops of a freshly prepared 5% solution of sodium nitropruside or simply add a pinch of freshly ground solid sodium nitropruside and 2ml of concentrated ammonium solution. Thoroughly mix the contents of the test tube and allow it to stand undisturbed for about 10minutes. The development of a characteristic permanganate or magenta colour will indicate the presence of acetone.

Method-II: Take about 2-3ml of urine sample in a test tube and add a few drops of 5% solution of sodium nitropruside, then add a dilute solution of KOH, little by little to make it alkaline. This should be tested from time to time by litmus paper during addition of KOH solution. The development of deep red colour will indicate the presence of acetone.

Note: It may be remembered here that a similar colour is also developed when creatinine is present in the sample. This could be verified by neutralizing the above mixture with acetic acid. If the colour persists, presence of acetone is confirmed.

Exercise No. 23

To detect the presence of blood in the sample of urine.

Theory: The blood can appear in the urine either in the form of RBCs or only haemoglobin. The presence of red cells is usually detected in urinary sediment by a microscopic examination. The presence of haemoglobin is due to haemolysis of RBCs. There are several tests for the detection of the presence of blood in the urine. For example, in addition to, microscopic examination for RBCs, they include stereoscopic examination for blood pigments and chemical tests. The best diagnostic test is the microscopic screening of urine sample. However, the following tests are commonly used in most of the pathological and class laboratories.

Test No. 1 Benzidine or Guaic Test for blood in urine

Take 2-3ml of urine sample in a test tube and add to it 2ml of benzidine solution. Mix it well and add 1ml of 3% solution of hydrogen peroxide. The development of a characteristic blue or bluish green colour will indicate the presence of blood in urine.

Note: Benzidine is carcinogenic, therefore avoid inhaling it and handling it. It should be used very cautiously.

Test No. 2 Haemin crystal Test for blood in urine

First of all concentrate the urine sample by drying it in hot air or keeping the sample in an oven set at appropriate temperature. Now place 1-2 drops of this concentrated urine sample at one end of a slide. Dry it over a low flame, then add a micro drop of normal salt solution, 1-2 drops of acetic acid and place a cover glass over it. Heat it over a low flame till effervescence. Immediately remove the slide and add, from side of coverslip, 1 or 2 drops of acetic acid. Allow it to cool down and examine your preparation. The formation of dark brown crystals of rhomboidal shape, of course, of different sizes will indicate the presence of blood in the sample.

Exercise No. 24

Microscopic examination of urinary sediment.

The sediment may contain crystals of various kinds, such as phosphates, oxalates, urates, uric acid and amino acids, like cystine. Epithelial cells from the lining of various regions of the urinary tract may also be found. The pus cells and casts are of special

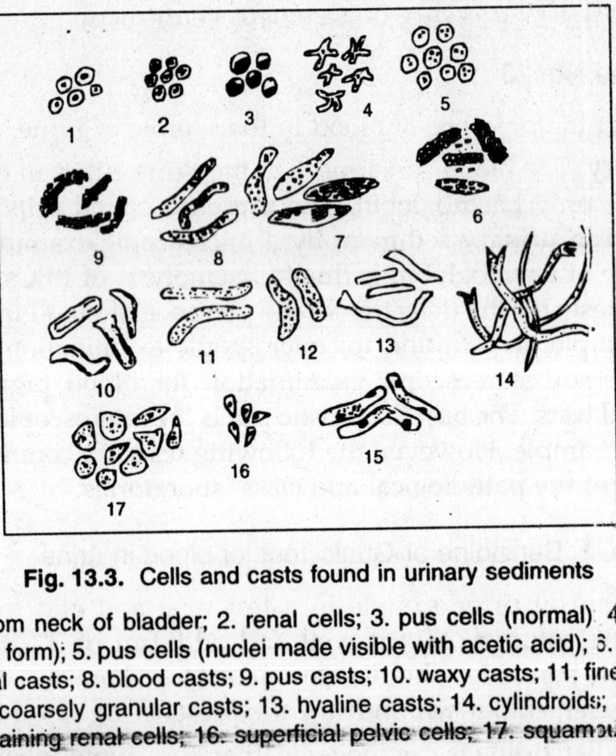

Fig. 13.3. Cells and casts found in urinary sediments

1. Cells from neck of bladder; 2. renal cells; 3. pus cells (normal) 4. pus cells (amoeboid form); 5. pus cells (nuclei made visible with acetic acid); 6. fatty casts; 7. epithelial casts; 8. blood casts; 9. pus casts; 10. waxy casts; 11. finely granular casts; 12. coarsely granular casts; 13. hyaline casts; 14. cylindroids; 15. hyalinic casts containing renal cells; 16. superficial pelvic cells; 17. squamous epithelial cells from the bladder.

significance as they help a physician in the diagnosis of certain kidney disorder. Occasionally the pus cells are found in normal urine. Their presence in large numbers, however, indicate the possibility of infection either in kidney or urinary tract or both. The presence of pus cells is known as *pyuria*. Pus cells are actually white blood cells and are readily distinguishable from other cells by their nuclei and peculiar granular appearance. Urine samples having pus cells normally give a positive test for albumin. The other materials of various kinds appear under microscope as cylinder with rounded ends and are called casts. Some casts are hyaline, others are granular, and still others are filled with pus cells, epithelial cells or fat globules. Each type of cast has its own particular significance. The presence of albumin together with cast indicates more serious condition than the presence of casts alone (Figure 13.3).

Frequently in a urinary sediment bodies called cylindroids, which are often mistaken for casts but which have no pathological significance, are observed. These are usually longer and narrower, they tend to be flat rather than rounded and frequently have branching ends.

14

Physiology of Respiration

INTRODUCTION

Respiration is a life process, which is defined as the gaseous exchange between an organism and its environment, may be air or water. The salient features are the intake of oxygen and the output of carbon dioxide. The gaseous exchange is easily facilitated in small organisms because they can take the oxygen directly from the air or water (in case of aquatic animals) which is carried in all their structures and give off carbon dioxide from all parts directly into the surrounding medium.

Measurement of oxygen consumption

A variety of methods are available for the measurement of oxygen consumption depending upon the resources and number of students.

For aquatic animals some kind of continuous flow system is required with a suitable outlet at some point in the system to allow the withdrawal of water for analysis. For terrestrial animals, the procedure adopted depends somewhat on size of the animal. The most popular method for small animals and pieces of tissues are probably the manometric methods of which the best known are the Warburg Manometers or Warburg and Dixon Barcroft manometers.

Exercise No. 1

Determination of the oxygen consumption of the larva of housefly *Musca domestica* by manometer technique.

Equipment and reagents required:

Warburg's manometer, constant temperature bath, 10% KOH solution, Brodies fluid, pipette (1ml), filter paper 1.5 x 1.5cm folded in accordion pleats.

Preparation of Brodies Fluid: Dissolve 23g of sodium chloride in 500ml of distilled water (density = 1.033).

Theory: Since many metabolic processes involve the exchange of gases, a large number of techniques based on manometer have been developed for use in biological laboratories. The underlying principle of manometer apparatus is based on the well-known gas law. At constant temperature and volume any change in the amount of gas in a system will produce a corresponding change in the pressure of that system. Best known manometric apparatus is the Warburg Respirometer designed by Otto Warburg in 1926.

Warburg Apparatus

This is widely used to measure the oxygen (O_2) uptake and carbon dioxide (CO_2) output by small animal or tissue. The apparatus (Figure 14.1) consists of detachable flask with a side arm attached to the arm of the manometer. This flask (reaction vessel) has a central well and a side arm equipped with a gas vent and solid ground glass stopper, a fluid reservoir with Brodies fluid of known density (1.033 in this case), a screw clamp, a three way stop cork and curved ground glass joint for attachment of reaction vessel. During the operation both the gas vent in reaction vessel and the three-way stop cork in manometer are closed, thus sealing the system. Flasks remain immersed in a constant temperature water bath prior to the start of an experiment. A reference point is selected on the closed side of manometer; the column height of the Brodies fluid in the open arm is also recorded. The flask remains immersed in a constant temperature water bath (Figure 14.2) and both flask and manometer are mechanically shaken.

The volume of air space in flask and capillary tube connected to it is kept constant by adjusting the level of fluid in the closed arm of the manometer to the same level before taking each reading. Only the pressure varies as the quantity of the gas in the flask changes. Therefore this is measured by the difference between the heights of the fluid in the two arms of manometer. Since both the temperature of water bath and atmospheric pressure may slightly vary during the course of experiment, it is necessary to have a control flask containing no animal but under the same condition as experimental flask. This is called a thermobarometer and as the name suggests serves to correct the fluctuations in the temperature of bath and atmospheric pressure. Therefore the pressure changes must be substracted from or added to (as the case may be in all the experimental settings).

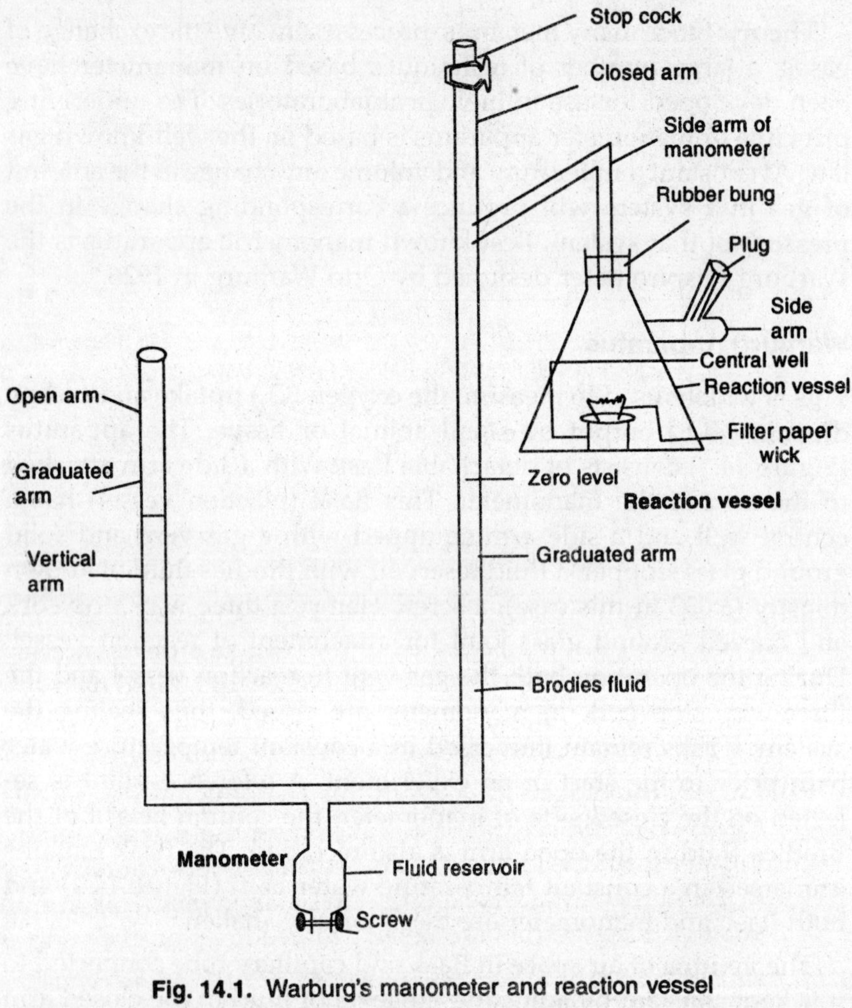

Fig. 14.1. Warburg's manometer and reaction vessel

The animal consumes O_2 and evolves CO_2. Simultaneously pressure changes in the flask containing an animal take place. The difference represents the net gaseous exchange. If CO_2 is absorbed by KOH as it is evolved, the pressure change represents O_2 consumption only. The difference between total gaseous exchange and O_2 consumption represents the CO_2 evolved.

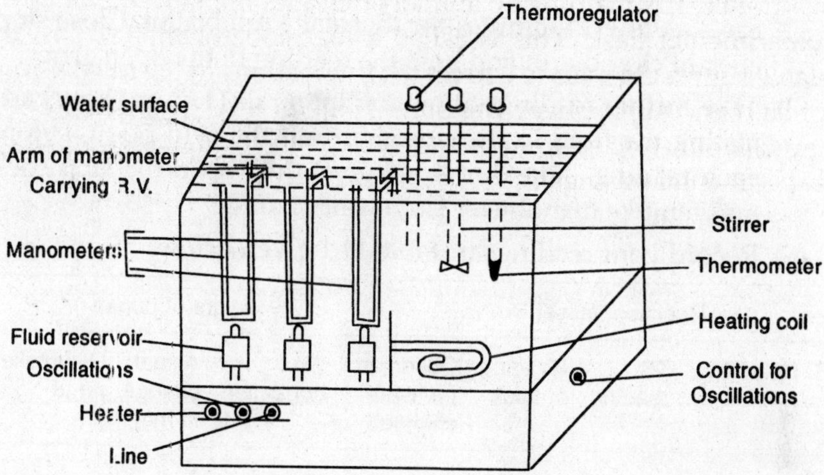

Fig. 14.2. Illustration of constant temperature water bath carrying manometers with reaction vessels (submerged in water)

Procedure

(1) Set the constant temperature water bath at 35°C.

(2) Place 5 flesh fly or blowfly or housefly larvae in the main chamber of the reaction vessel or flask.

(3) Pipette 0.2ml of 10% KOH solution into the central well. Put a folded filter paper piece in the central well in order to increase the surface area, for the absorption of CO_2 liberated by respiration of larvae. Also leave a large surface of filter paper above the top of the well to further increase the surface area.

(4) Prepare three more flasks with five larvae in each and KOH absorbent in the central flask. Take one or more flask containing all except the larvae. This is used as a control and treated in the same way as an experimental manometer. This in true sense acts as thermobarometer. The readings on the control manometer register changes in temperature or barometric pressure and must be substracted from those taken from experimental manometer.

(5) Attach the flask to manometer side arm and secure them with spring clips. See that the flask is properly sealed to the ground

glass end of the manometer arm, using vaseline, making sure that the manometer is not clogged with vaseline.

(6) Place all the manometers in the water bath. Adjust the fluid (stop cock open) in the closed arm at zero point (15 in this case). Allow 15 minutes for thermal equilibrium. Close stop cock of closed arm and note the reading in the open arm.

(7) Take further readings at regular intervals (1/2 or 1hour) adjusting the fluid in the closed arm to the zero point, before each reading. Take 3 or 4 readings then open the stop cock and remove manometer from water bath.

(8) Record your readings in Table 14.1 given below:

Reaction Vessel No.........					k= weight of pupae		
S. No.	Time	TB reading	Reading of flask	Observed pressure changed	TB correction	Actual change in mm	O_2 uptake on µl
1.							
2.							
3.							

Calculation: The larvae consume O_2 and evolve CO_2. Simultaneously, pressure changes in the flask containing the larvae represent the net gaseous exchange, when CO_2 is absorbed by KOH as it is liberated. The pressure change represents O_2 consumption only. The difference between total gaseous exchange and O_2 consumption represent the CO_2 evolved.

Correct the pressure changes in the experimental flask from thermobarometer reading to find out the real pressure change (h) in each of the experimental flask:

Formula – 1

$$X = \frac{h[Vg\ 273 + Vfa]\ hk}{T/P_0}$$

Formula – 2

$$X = \frac{h(Vg\ 273/T)}{P_0}$$

Where,

X = amount of gas consumed (in litre)

h = observed pressure change in the manometer reading in nm (actual pressure exchange)

Vg = volume of gas phase in the flask and manometer upto zero point

T = temperature of water bath in degree absolute 'A' (273+°C)

P_o = normal atmospheric pressure 760mm (Hg) expressed as mm of manometer fluid, thus

$$P_0 = \frac{760 \text{ specific gravity of Hg } 13.6}{\text{specific gravity of manometer fluid at } 0.33} = 10{,}000$$

Vf = volume of fluid in vessel (main chamber)=10,000

α = solubility in liquid in vessel of gas evolved (as ml gas per ml) liquid at STP

K = Flask constant (fundamentally this consists of calibrating the system that from the observed pressure change one can calculate the amount of gas evolved in μ lit or microlitre).

In the above formula Vg, T, & Po are known for a given experiment and are constant for a given experiment. These values determine the flask constant K

Hence, X = hK

Now, from the above observed pressure change h, the amount of gas consumed can be calculated very easily by simply knowing h and K. By conviction the gas produced or consumed is given a positive or negative sign, respectively.

Result

(1) Average O_2 uptake μl/mg/h

(2) Average CO_2 uptake μl/mg/h

Precautions

(1) Manometer should be handled very carefully since they are expensive.

(2) All the joints should be airtight.

(3) Manometer must be equilibrated with the temperature of water bath.

(4) The stop cock should be opened before removing the manometer from the water bath otherwise fluid will be sucked into the flask.

Limitations

Experimental animal respire in an environment from which the O_2 is being constantly utilized.

(1) First of all keep the manometer for 15 minutes in water bath for temperature equilibrium.

(2) After this, three-way stopcock is closed.

(3) The fluid level is adjusted and brought at 15cm and the changes are read in open arm.

For this there are three conditions.

(a) Rise in T.B. fall in RV – add

(b) Fall in T.B. fall in RV – substract

(c) Rise in T.B. rise in RV – substract

RV reading from that of T.B. reading.

Where, T.B.= Thermobarometer; RV = Reaction Vessel

Exercise No. 2

Demonstration of difference in oxygen consumption between species and between the individuals of a species.

Animals, equipments & chemicals required: Frogs, toads, lizards, chicks, mice, rats, *etc.*; glass jar with wide mouth, rubber corks, tubing to fit in the mouth of the jar. Laboratory stands, metal gauze or flywire, glass tubings, rubber tubing, clamps for rubber tubing and soda lime.

Theory: The underlying principle is that animals consume oxygen. The oxygen used is combined with hydrogen, obtained from the oxidation of foodstuff to form water. During this consumption there is an associated release of the energy originally contained in the foodstuff as potential energy. The energy released is immediately trapped and transferred to the energy store of the animal.

The harder the animal works, the more energy is required and so the more oxygen. For example, a man resting uses approximately 0.2ml O_2/g/h, whereas a man working uses nearly 4ml O_2/g/h.

The variations in oxygen consumption are related with: (i) temperature, (ii) nutrition, (iii) body weight, (iv) stage in the life cycle, (v) sex, and (vi) hormonal and environmental effects.

The differences in oxygen consumption, referred above, are differences between individuals of one species, but of course different species also consume different amount of oxygen.

Apparatus: Two cylindrical cans of tin, one of which should be of such a dimension both in diameter and length that it closely fits inside the other one. The tin can should be open at one end. The larger can should have a piece of metal tubing soldered through a hole in the centre of the bottom so that the tube extends (inside) the length of larger tin, whereas on the other side of the bottom of larger can, the tube should remain projected about 1 ½ to 2" outside the tin can. The smaller can should have a strip of centimetre graph paper pasted on the outside and a hook soldered on the outside of the bottom or closed ends (Fig. 10.3).

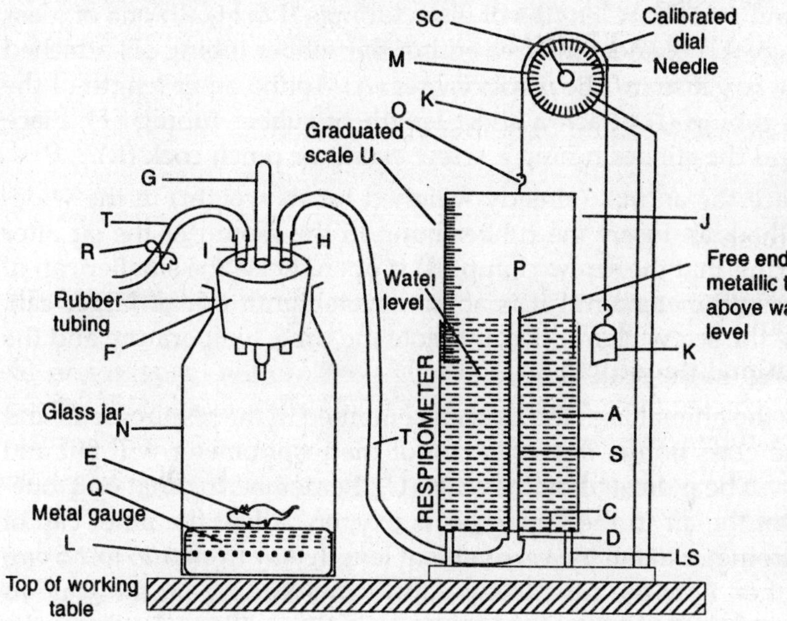

Fig. 14.3. Assembly for the measurement of oxygen consumption of small animals.

Procedure

Set up the assembly as shown in Figure 14.3. Take the larger can (A) and fill it with water and rest it on the working table. Moisten the projected outer end of the metal tube (D) and push on to it a length of rubber tubing (T). Fit the smaller can (J) into the larger can and allow the air trapped (in P) to escape through the rubber tubing. The smaller can will come to rest almost completely enclosed by the larger one. Tie a piece of non-elastic surgical thread (K) on to the hook (O), at the centre base of the smaller tin can. Drop the thread over the pulley (M) and attach other end of thread with counter weight (K). The weight should be such that the smaller tin just sinks through the water contained in larger can, thus expelling air contained in P. The respirometer should be placed horizontally on its stand as ascertained by levelling screws (LS) and spirit level.

Place soda lime (L) in the bottom of the wide mouthed glass bottle (N). Arrange a piece of metal gauze (Q) above the soda lime so that the experimental animal (E) does not come in contact with the soda lime. Place the rubber bung (F) with three holes and push through each of its hole (moistened or greased first) a thermometer (G) and two short lengths of glass tubings (I & H). To one of glass tubings (H) connect the free end of the rubber tubing (T) attached to the larger tin of the respirometer (A). To the other length of the glass tubing (I) attach a short length of rubber tubing (T). Place around the rubber tubing a screw clamp or pinch cock (R).

Place the animal (already weighed for its weight) in the wide-mouthed jar. Insert the rubber bung in the mouth of the jar after ensuring that the screw clamp (R) is open. Raise the smaller can of the respirometer until it is about seven-eighth out of larger can. Close the screw clamp (R), and note the time, temperature and the reading on the scale (U).

As the animal uses the oxygen contained in the respirometer and in the glass bottle, the inner can of the respirometer will fall and this will be indicated on the scale (U). Remembering that only one-fifth of the air in the apparatus is oxygen. Allow the inner can of respirometer to fall by a convenient length (say from 0.25 to 2.5 cm) and then note the time, temperature and reading on the scale as accurately as possible. Open the screw clamp (R) and remove the rubber tubing (T) so that a fresh supply of air is available to the experimental animal.

After an interval of short duration (say 15-30 minutes), repeat the experiment to obtain duplicate readings: Then perform a similar experiment with one animal of the same species and two animals of different species.

Sample, calculation and results

Let us presume that the

(i) Inner can fall in 15 min h = 1.1 cm

(ii) Diameter of the tin can J = 6.0 cm, hence the radius r = 3 cm

Volume of the oxygen needed $\pi r^2 h = 3.14 \times (3 \times 3) \times 1.1$ ml or

$3.14 \times 9 \times 1.1$ ml

weight of animal 65.7 g

$$O_2 \text{ consumed per hour per gm.} = \frac{3.14 \times 9 \times 1.1 \times 60 \text{ ml}}{65.7 \times 15}$$

$= 1.9$ ml

Result: Rate of oxygen consumption is 1.9 ml/hr.

General Observations: It may be noticed that small animals use more oxygen per unit weight than larger ones.

Precautions

If after several minutes, no oxygen appears to have been consumed, check up the assembly as under:

(1) Is the counter balance weight K too heavy? Is the cotton thread stuck at pulley? Both of these points will prevent the small can from sinking downwards in the larger can?

(2) Is there any water trapped in the rubber tube connecting the respirometer with the glass bottle-containing animal?

(3) Is air leaking into the apparatus? If so, try greasing the suspected areas with petroleum jelly?

(4) Is the soda lime fresh? The purpose of the soda lime is to absorb the CO_2 evolved or breathed out by the animal?

(5) Is the animal under investigation habitual to stop breathing for short periods?

Note: Similar type of assembly can be used to experimentally ascertain, the effect of (a) temperature, (b) sex, (c) light, *etc.* on the rate of oxygen consumption of individuals of equal weight of the same species?

Exercise No. 3

Measurement of the lung volumes using respirometer.

Theory: This means the vital capacity (V.C.) of lungs, *i.e.* the total volume of the air that can be accommodated during inhalation and expiration. This is equivalent to the sum of tidal air + supplemental air and complimental air. Information pertaining to this issue is very helpful in assessing the actual state of the functioning of lungs as well as the general health status of an individual. Both these informations are useful to a surgeon for putting a patient on the operation table. This is measured with the help of an apparatus called ***spirometer.***

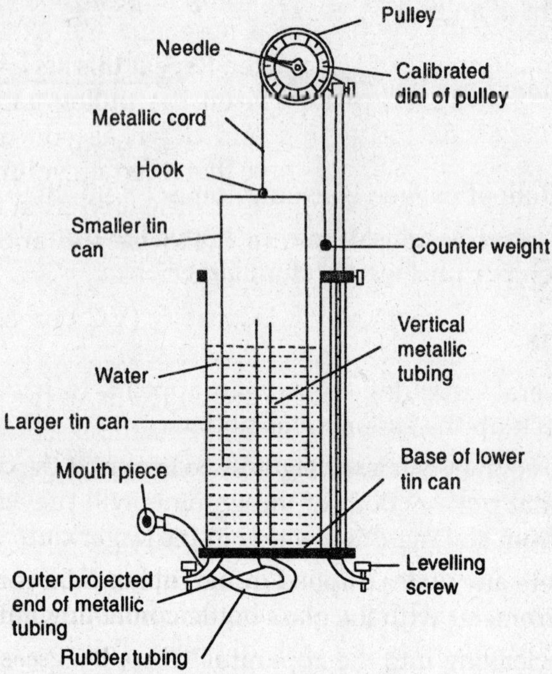

Fig. 14.4. The respirometer measuring lung volumes.

Apparatus: The respirometer (Figure 14.4.) consists of a large cylinder tin can of adequate capacity. This cylinder is provided with a vertical metallic tubing soldered at its base in such a way that the greater length projects vertically inside the cylinder while a small length projects outside from the centre of the base. Sufficient quantity of water is poured in the larger can, so that the level of water remains just below the top end of the vertical tube. Another small cylindrical tin can, being a close fit inside the larger one in an inverted condition acts as an inverted floating close fit. It is kept over and inside the larger cylinder with the help of a light metallic cord running over a pulley. The chain is attached at one end to a hook placed at the centre of the outside base of the inverted smaller cylinder and on the other end of the counter weight. The float is kept in balanced condition throughout its movement. The dial on pulley is so calibrated as to directly indicate the volume of the air.

Procedure

Sterilize the mouthpiece with a piece of cotton swab wetted with alcohol. Keep the apparatus in a perfect horizontal position by operating the leveling screws and using a spirit level. Block the passage of nose with the help of a nose clip and exhale the air through the mouthpiece of spirometer. Repeat this several times to have a good practice. Keep an eye on the pulley and float air. Observe that the float moves up and down as you exhale and inhale respectively and with this note that a certain volume of air is passed out and taken in. This volume is measured from the graduated pulley, which keep on rotating clockwise and anticlockwise depending upon air exhaled and inhaled.

For actual measurement of Vital Capacity (V.C.) proceed as under:

(1) **Tidal Air (TA)** Adjust the cylinder at 0 mark level on dial. Breathe in the normal way through the mouthpiece and note down the volume of air inspired and expired. No extra pressure should be used.

(2) **Supplemental Air (SA)** When you have expired for the TA in normal way, then exhale through the mouthpiece as much air as you can into the spirometer and note down the volume of air as recorded on the scale of pulley. This much volume is equivalent to the supplemental air or expiratory reserve volume (ERV).

(3) **Complimental Air (CA)** After a normal inspiration as for the total air, breath as much air as you can by deeply and forcefully breathing the air. This is equivalent to complimental air or inspiratory reserve volume (IRV)

Calculation: From the measurement made above, the vital capacity (VC) or total lung volume (L.V.) can be worked out as under:

$$VC = TA + SA + CA$$

Precautions

(1) The mouthpiece should always be sterilized before either being used afresh or by any another student.

(2) Do not watch yourself at the scale during experiment, better ask your partner to note down the lung volume as indicated on pulley.

(3) Always drain off the water after the experiment.

Exercise No. 4

Measurement of Basal Metabolic Rate using Benedict's Roth Spirometer

Theory: The energy expended by the animal, when it is at complete rest is assumed to represent the energy required to maintain the animal in proper condition. In homeotherms this level of metabolism is known as Basal Metabolic Rate (BMR) and in poikilotherms as the Standard Metabolism.

In practice the conditions of complete rest are unobtainable, hence BMR is measured well after a meal and during sleep.

For the determination of BMR, the measurements of the rate of oxygen intake and carbon dioxide output are made only when a person: (1) is at complete rest, *i.e.* when the mind and body are free from mental or physical stress and strains, respectively, and (2) the person has not taken meals at least 12-15hours before putting him on test. The commonest way of expressing basal metabolism is in terms of heat production, the unit used being Kcal. Heat production is normally calculated from oxygen consumption.

From the equations describing oxidation of major classes of foodstuffs, it is possible to calculate the amount of heat equivalent to the utilization of one litre of oxygen for each of them. These figures amount to 5.05Kcal for carbohydrate, 4.69 Kcal for fats and 4.25Kcal for protein. These figures are sufficiently close for a single figure to be used for an animal with omnivorous diet for example, 4.8 is used for man. If the food intake of an animal is known precisely and particularly if it uses overwhelmingly one type of food, the accuracy is greater.

Apparatus: In usual practice under laboratory conditions the BMR is measured by means of a Benedicts Roth Recording spirometer (Figure 14.5). This spirometer is a slightly modified form of the ordinary used simple spirometer. It consists of two metallic cylindrical jars, one fitting into the other. One of these acts as a float. The capacity of the float is approximately 6000ml. At its base on the outer side it is fitted with a hook, to which is attached a chain that passes over a pulley fixed on a stand. The float is counter balanced by a weight that is attached to the other end of the chain. A thermometer is fitted on the float for recording internal temperature. Up and down movements of the float are recorded by means of a gravity writing ink pen adjustable by a lever at the base of the counter balance. The ink pen leaves marks on the paper mounted

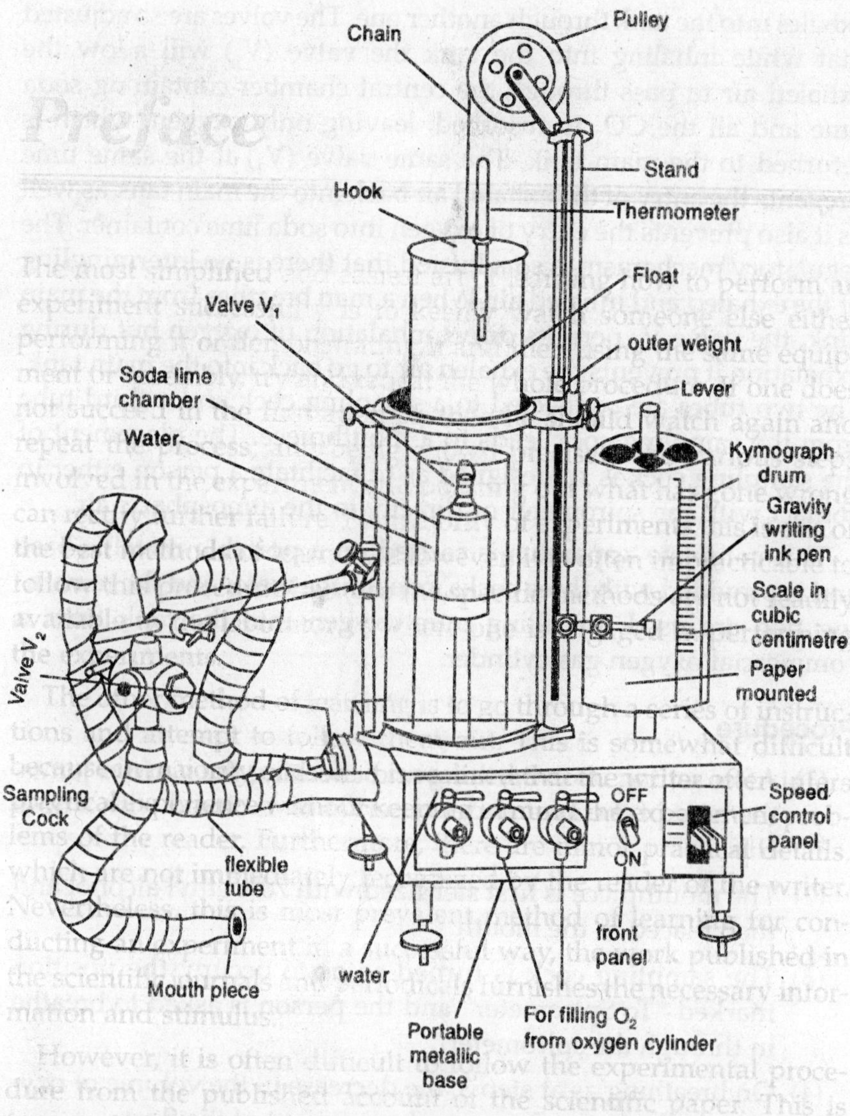

Fig. 14.5. Benedict Roth Spirometer.

on kymograph drum when the float moves downwards. The recording metallic drum is of 12 x 6 inches dimensions and is of the lift type. The drum is driven by an electrically operated motor, workable on 2 or 4 adjustable speeds. The papers used on the drum are of specially glazed type. The speed is adjusted through a lever located on the front panel. The spirometer is equipped with two flexible tubes and the sampling cock in such a way that a person under experiment inhales from the tank through one tube and exhales into the tank through another one. The valves are so adjusted that while inhaling into the tank the valve (V_1) will allow the exhaled air to pass through the central chamber containing soda lime and all the CO_2 is absorbed; leaving only oxygen, which is returned to the main tank. The same valve (V_1) at the same time prevents the entry of the exhaled air back into the main tank as well as it also prevents the entry of oxygen into soda lime container. The regulatory mechanism is so adjusted that there is no intermingling of the exhaled and inhaled air. When a man breathes from the main tank, the valve V_2 permits direct inhalation of oxygen but during exhalation it prevents the exhaled air to go back into the main tank. The two tubes are connected to a sampling cock and a third tube from the sampling cock leads to a mouthpiece. The placement of the sampling cock is so designed as to facilitate a person either in contact with the spirometer or directly to the atmospheric air.

This complete apparatus is carried on a portable metallic base and is provided with three cocks, one for the water and the remaining two are used for filling pure oxygen into the tank from a commercial oxygen gas cylinder.

Procedure

(1) After setting up everything and checking the various joints, the nose of the man under observation is closed with a nose clip.

(2) The mouthpiece is first sterilized with 70% ethyl alcohol and then placed in the mouth.

(3) The sampling cock is turned so as to occupy the position marked "To Spirometer" and the person is asked to breathe in through the spirometer.

(4) On breathing as at step 3, the decrease in the volume of oxygen results in the downward movement of the float.

(5) From the reading at step 4, the rate of oxygen consumption per unit time is worked out.

(6) When both the total amount of oxygen consumed and carbon dioxide produced are measured, then the respiratory quotient (RQ) is calculated by the formula

$$RQ = \frac{CO_2}{O_2}$$

The disadvantage of this method, which is known as indirect calorimetry, is that it assumes that metabolism is completely aerobic.

Precautions

(1) The person being examined should avoid breakfast and his stomach should be empty.

(2) The person should be advised to lie down on a smooth bed in a comfortable position for at least 20-30 minutes before being examined.

(3) The temperature of the room should be almost normal and the person should be asked to cover himself with a blanket preferably woolen in winter season.

(4) The atmosphere of the laboratory should be kept quiet and noiseless.

(5) The experiment should be repeated at least twice with a minimum interval of 10-15 minutes.

(6) Each time determine the rate of oxygen consumption separately.

Method for Calculation of BMR

For calculating BMR, the most common method is that of DuBois. The main factor involved in this method is the use of body surface area, which is regarded as the chief factor related with the heat production and its elimination.

The body surface area of a person is calculated by the formula given below:

Approx. area in sq. cm = Height in cm × 0.725 × weight in Kg × 0.425.

Presuming an RQ of 0.82 and oxygen used at this RQ producing 4.8calories of heat, calculate the heat produced per hour per square metre of body surface.

For calculation it should be remembered that for a foodstuffs consisting of carbohydrate, proteins and fats, the RQ is 0.82 and for every litre of oxygen consumed 4.8 calories of heat are produced. That means, if only oxygen consumption is measured, then the heat produced can be calculated per hour per square metre of body surface. After calculating the BMR of a person, the results can be verified from table 14.2.

Table 14.2: Age wise average BMR of normal human being.

Age of a person		BMR in Kcal/ sq. metre/ hour	
		Female	Male
	2	52.5	57.0
Gr. A	6	50.6	53.0
	8	47.0	51.8
	10	45.9	48.5
Gr. B	16	38.8	45.7
	20	36.1	41.4
	30	35.7	39.3
Gr. C	40	35.7	38.0
	50	34.0	36.7
Gr. D	60	32.6	35.5
	60 above		

Note: The RQ is different for different kinds of food, e.g. RQ for aerobic combustion of hexose sugar, the quantity of CO_2 formed is equal to quantity of O_2 utilized:

$$C_6H_{12}O_6 + 6O_2 = 6CO_2 + 6H_2O$$

Here, the ratio (Volume of CO_2 formed/ Volume of O_2 utilized) is equal to unity or 1, for proteins 0.8 and for fats 0.7. For mixed foodstuffs, (which is the usual diet) RQ is 0.82

Exercise No. 5

To measure the oxygen consumption by small animals using simple apparatus.

Apparatus: A respiratory chamber (either a wide mouthed glass bottle or desiccator with open lid), a manometer, a chemical balance, rubber bung, glass and rubber tubings, syringe, pinch cork.

Reagents: 10% KOH solution, filter paper.

Animals: Rat, mice, larvae of insects, e.g. silkworm, cockroach, beetles, *etc.*

Procedure

The animal is weighed and is sealed inside the chamber (Figures 14.6 a, b & 14.7). Enough of oxygen is blown in to give 50% enrichment and the circuit is closed to the outside and connected to a manometer. After a 30minute equilibration period, the zero reading on the manometer is taken. Further readings are taken 1, 2 and 3hours later. The CO_2 is absorbed by soda lime in the respiratory chamber. The volume of the chamber should be known and the fall in pressure as read in mm Hg from manometer and the duration T from the timepiece. All these values are entered in the formula given below for:

O_2 consumption in ml/hr/ unit body weight

$$= V \times \frac{P}{760} \times \frac{273}{273+t} \times \frac{1}{T}$$

Generally speaking, the rate of oxygen uptake of the egg is much lower than in the adult, but is high in the larval stages, especially before moulting and falls again in the pupal stage. The oxygen consumption rate of adult blowfly *Pormia regina* at rest is about 1.75mm^3 of O_2 per hour per mg of live weight.

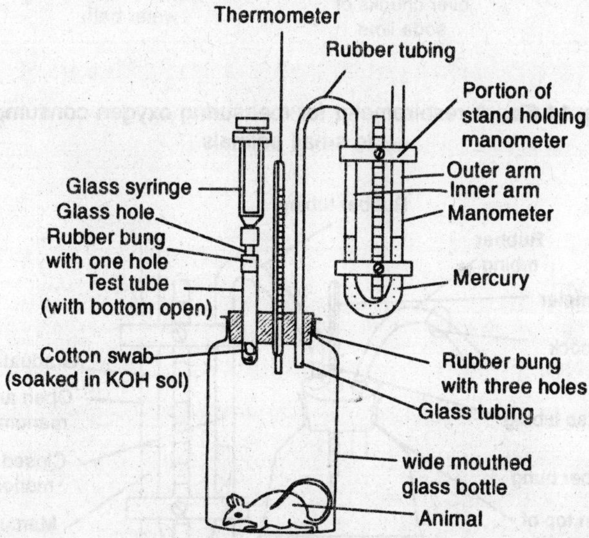

Fig. 14.6a. Assembly for measurement of oxygen consumption of rat or other small animals

For much large animals, e.g. cattle, there are modern respiratory chambers as at the Hannah Research Institute, Agra. The parameters measured include oxygen consumption, carbon dioxide production, methane production, urinary nitrogen, food consumption and excretion.

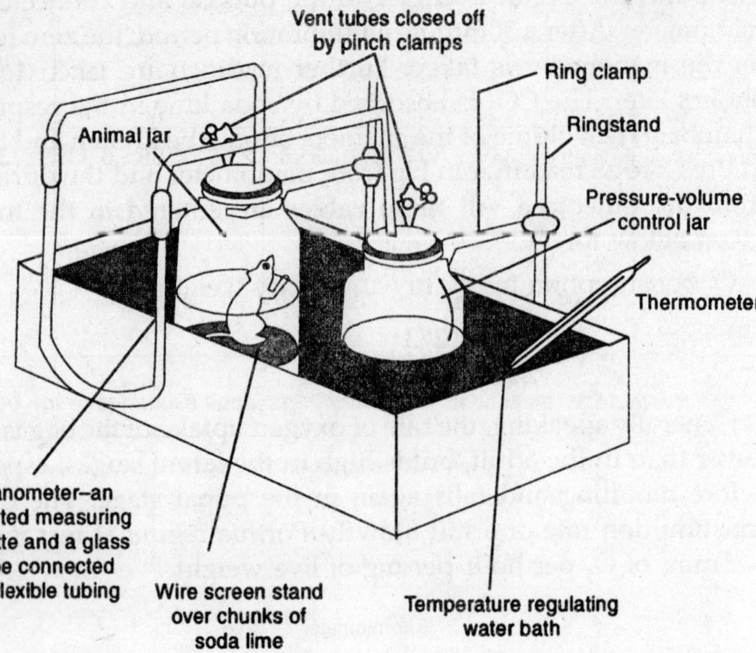

Fig. 14.6b. A respirometer for measuring oxygen consumption in small animals

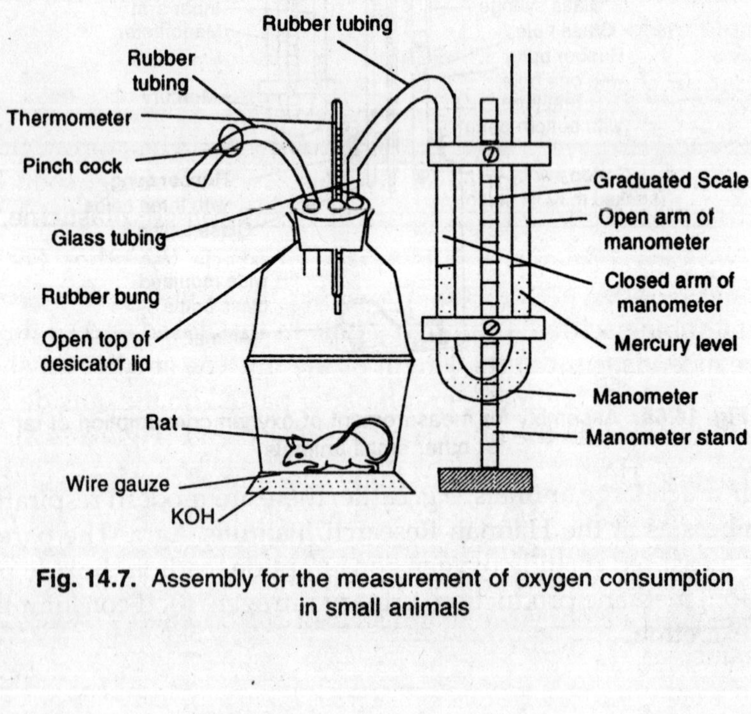

Fig. 14.7. Assembly for the measurement of oxygen consumption in small animals

Table 14.2: Oxygen consumption during development of *Drosophila melanogaster*

Age in hours	25	40	47	60	90	98	110	130	150	180
Oxygen uptake in mm^3/hr by one insect	3.46	4.68	3.60	2.52	3.96	2.16	0.86	0.93	0.73	1.80
Developmental stage	first instar	Second instar		Third instar			Pupa		Emergence of adult	

Table 14.3: Seasonal differences in gaseous exchange in the frog

		Respiratory exchange (ml/Kg/hr)					
Time	Wt.	Cutaneous		Pulmonary		Total	
		O_2	CO_2	O_2	CO_2	O_2	CO_2
Spring	46g	51	145	160	70	211	215
Autumn	54g	54	92	51	15	105	107

Exercise No. 6

To measure the oxygen consumption by the whole animal using a simple method.

Apparatus and chemicals

Glass jar with a wide mouth, rubber tubing to fit, laboratory stands, glass tubing, metal gauge or fly wire, rubber tubing, soda lime, toads, frogs, lizards, mice, squirrels, cotton thread, plasticine, etc.

Two cans, one of which is a close fit inside the other. Each tin should be open at one end. The larger tin should have a piece of metal tubing soldered through a hole in the closed end so that the tube extends (inside) the length of the tin. The smaller tin should have a strip of centimetre graph paper pasted on the outside and a hook soldered on the outside of the closed end (Figure 14.8).

Theory: The oxygen taken in by the animal combines with hydrogen obtained from the oxidation of food stuffs to form water. During this combination there is an associated release of the energy originally contained in the foodstuffs. The energy released is immediately trapped and transferred to the energy store of the animal.

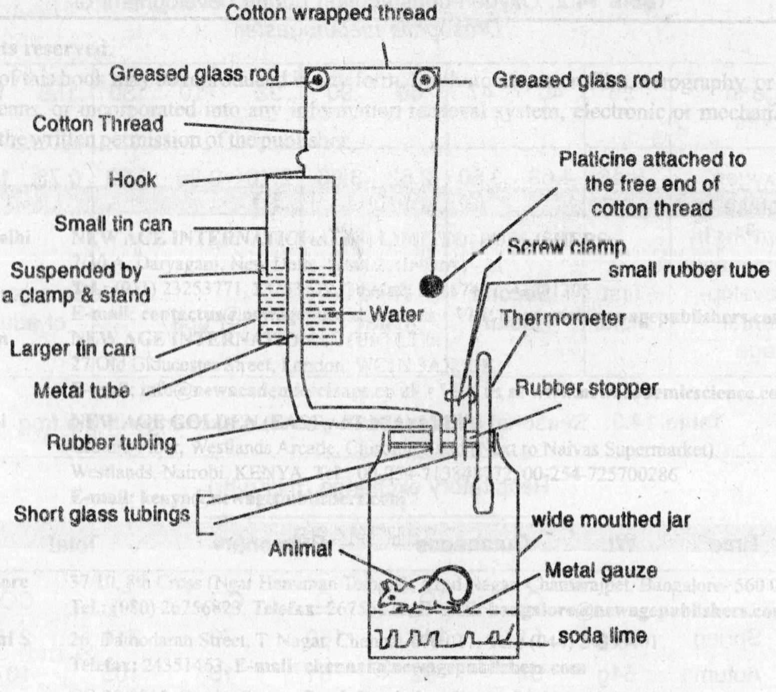

Fig. 14.8. Diagrammatic representation of the respirometer

The harder an animal works, the more energy is required and so the more oxygen. For example, a man resting uses approximately 0.2ml O_2/ g/h whereas a man working uses nearly 4ml O_2/ g/h. Oxygen consumption also varies accordingly to (a) temperature, (b) nutrition, (c) body weight, (d) stage in life cycle, (e) sex, and (g) hormonal and environmental effects. Such differences in oxygen consumption do exist between individuals of the one species, but, of course, different species also consume different amounts of oxygen. This is mostly due to the differences in the biochemical make up between species. The sole aim behind this experiment is to give a better understanding with regard to the differences in oxygen consumption between species and between individuals of species simplified respirometer.

Procedure

First of all set up the apparatus as shown in Figure 14.8. Fill the larger tin can with water and suspend it by means of a clamp and laboratory stand about 18″ above the bench. Moisten the metal tube and put onto it a length of rubber tubing. Now place the smaller tin can (carrying a centimetre graph paper pasted on its outer side), into the larger and allow the air trapped in space to escape out through the rubber tubing. After a while the smaller tin can will be almost completely enclosed by the larger one. Now tie a piece of cotton thread onto the hook at the closed outer end of the small tin can. Drop the cotton thread over to lightly greased glass rods and attach plasticine to the other end of the cotton thread. The weight of the plasticine should be such that the smaller tin just sinks through the water, thus slowly expelling the air in the space.

Fill the bottom of the wide mouthed jar with soda lime. Arrange a piece of metal gauge above the soda lime so that the experimental animal does not get a chance to come in its contact through three holes in the rubber tubing of the size of the mouth of glass jar. Push moistened two glass tubing. To one of the ends of the glass tubing connect the free end of the rubber tubing attached to the larger tin of the respirometer. Attach a short length of rubber tubing to the end of other glass tubing. Place around the rubber tubing a screw clamp. In the third hole in rubber stopper insert a thermometer.

Place the animal (or animals) in a wide mouthed jar. Insert the rubber stopper in the mouth of the jar after making sure that the screw clamp is open. Rinse the smaller can of the spirometer until it is about seven-eighths out of the larger can. Close the screw clamp, and note the time, the temperature and the reading on the scale. As the animal uses the oxygen contained in the respirometer and in the glass jar the inner can of the respirometer will fall and this will be indicated on the scale. Remembering that only 1/5th of the air in the apparatus is oxygen, allow the inner can of the respirometer to fall by a convenient length (say, from 0.25 to 2.5cm) and then note the time, temperature and reading on the scale as accurately as possible. Open the screw clamp and remove the rubber stopper so that a fresh supply of air is available to the animal.

The experiment should be repeated to obtain duplicate readings. Then experiment with (1) animal of the same species but of different weight, and (2) of different species.

Calculation and Result

Fall in level of inner can in 15min	= 1.1cm
Diameter of the inner can	= 6.0cm, thus, r = 3.0cm
Volume of oxygen used	$\pi r^2 h$
	= $3.14 \times 9 \times 1.1$

Weight of animal 65.7g

$$O_2 \text{ consumed per hr per gm} = \frac{3.14 \times 9 \times 1.1 \times 60}{65.7 \times 15} = 1.9 \text{ ml}$$

Precautions

If no oxygen consumption is noticed then check the following:

(a) The weight of counter balance should not be too heavy. Cotton should not stuck at any point of application.

(b) There should be no tapping of water in the rubber tubing linking the respirometer with the glass jar containing animal.

(c) The apparatus should be kept air tight in case of leakage at any point apply grease at suspected points with petroleum jelly or put the glass jar under water.

(d) Fresh soda lime should always be used. The object of the soda lime is to absorb the CO_2 produced by respiration.

Suggested Further Reading

Abdridge, W.N. 1961. Biochemists Handbook. Ed. Long, C. Spon London

Adams, C.W.M.. Ed. 1965. Neurochemistry. Elsevier, Amsterdam.

Arnow, L.E. and Reitz, H.C. 1943. Introduction to Organic and Biological Chemistry. C.V. Mostry Co. St. Louis.

Armstrong, E.F. and Armstrong, K.F. 1934. The Carbohydrates. 5th Edn. Longsmann Green & Co., New York.

Anderson, A.K. 1953. Essentials of Physiological Chemistry. Toppan, Japan.

Baldwin, E. 1959. Dynamic aspects of Biochemistry. Harper and Row, New York.

Baldwin, E. 1963. Dynamic Biochemistry, 4th edn. Cambridge University Press.

Baker, J.R. 1959. Principles of Biological Microtechnique. Methuen, John Wiley, London.

Barka, T. and Anderson, P.J. 1963. Histochemistry. Harper and Row, New York.

Bancroft, J.D. 1975. Histochemical Technique. 2nd edn. London, Butterworth.

Bancroft, J.D. and Stevens, A. 1982. Theory and Practice of Histological Techniques. Churchill Livingston.

Bell, G.H., Davidson, J.N. and Scarborough, H. 1968. Textbook of Physiology and Biochemistry. English Language Book Society.

Bell, G.H. and Davidson, J.N. 1972. Textbook of Physiology and Biochemistry. E.S. Livingston, Edenborough.

Benchard, S. 1968. The Structure and Function of Enzymes. 1968. Benjamin, New York.

Bennet, T.P. and Frieden, E. 1966. Modern Topics in Biochemistry. Macmillan, New York.

Bergmeyer, H.U. 1963. Methods of Enzymatic Analysis. Verlag Chemie and Academic Press, New York.

Bier, M. 1959. Electrophoresis, Theories, Methods and Applications. Academic Press, New York and London.

Biggs, R. and MacFarlane, R.G. 1962. Human Blood Coagulation and Its Disorders. Blackwell, Oxford.

Biggs, R. and MacFarlane, R.G. 1963. Human Blood Coagulation and Its Disorders. 3rd edn. Blackwell, Oxford.

Block, R.J., Durrum, E.L. and Zweig, G. 1955. A Manual of Paper Chromatography and Paper Electrophoresis. Academic Press, New York.

Bobitt, J.M. 1963. Thin Layer Chromatography. Reinhold Publishing Corporation, Chapman & Hall, London.

Bodman, J. 1960. Chromatographic and Electrophoretic Techniques. Edited 1. Smith William Heinemann Medical Books, 2nd edn.

Bourne, G.H. 1960. The Structure and Function of Muscle. Three Volumes, Academic Press Inc., New York.

Boyr, P.D., Lardy, H. and Myrback, K. 1959. The Enzymes. Academic Press, New York.

Brown, G.L. 1968. Introduction to Physical Chemistry. Longmans Green, London

Bull, H.B. 1937. The Biochemistry of the Lipids. John Wiley & sons. New York.

Burstone, M.S. 1962. Enzyme Histochemistry and its Application in the study of Neoplasm. New York Academic Press.

Carleton, H.M. and Leach, E.H. 1947. Histological Techniques. Oxford University Press. New York.

Chayen, J. 1952. The Methyl Green – Pyronin Method. Exptl. Cell Res., 3, 652-655.

Clark, W.M. 1928. The Determination of Hydrogen Ion. 3rd edn. William and Wilkins Co., Baltimore.

Cole, C.F. 1967. Handbook of Physiology. American Physiological Society.

Cowdry, E.W. 1952. Laboratory Techniques in Biology and Medicine. 3rd edn. William & Wilkins, Baltimore.

Cramer, F. 1954. Paper Chromatography. Macmillan, London.

Crout, J.R. 1961. Standard Methods of Clinical Chemistry, edited David Selegson, Vol. II. Academic Press, New York.

Culling, C.E.A. 1974. Handbook of Histopathological and Histo-chemical Technique. London, Butterworth.

Dacie, J.V. and Lewis, S.M. 1963. Practical Haematology. 3rd edn. Churchill, London.

Dauson, H. 1964. A Textbook of General Physiology. J. & A. Churchill Ltd. London.

Davidson, J.N. 1969. Biochemistry of Nucleic Acids. Methuen, London.

Dill, D.B., Adolf, E.F. and Wilber, C.G. 1964. Handbook of Physiology. American Physiological Society, Washington.

Dixon, M. 1951. Manometric Methods. 3rd edn. Cambridge University Press, London.

Dixon, M. and Webb, E.C. 1964. Enzymes. 2nd edn. Longmans, London.

Faires, R.A. and Parks, B.H. 1958. Radioisotope Laboratory Techniques. George Newnes Limited, London

Florey, E. 1966. General and Comparative Animal Physiology. Saunders, Philadelphia.

Fruton, J.S. and Simmonds, S. 1958. General Biochemistry. John Willey & Sons, New York.

Gergely, J. 1964. Biochemistry of Muscular Contraction. Churchill, London.

Glick, D. 1962. Quantative Chemical Techniques of Histo- and Cytochemsitry. Vol. I. Wiley, New York.

Glick, D. 1963. Quantative Chemical Techniques of Histo- and Cytochemsitry. Vol. II. Wiley, New York.

Gomori, G. 1952. Microscopic Histochemistry. University Press, Chicago.

Gortner, R.A., Gortner, R.A. Jr. and Gortner, W.A. 1949. Outlines of Biochemistry. 3rd edn. John Wiley & Sons, New York.

Gradwohl, R.B.H. 1956. Clinical Laboratory Methods and Diagnosis. Vol. II. St. Louis: C.V. Mosby.

Gray, P. 1954. The Microtomit's Formulary and Guide. Balkiston. New York,

Gray, P. 1958. Handbook of Basic Microtechniques. New York, McGraw.

Gurr, E. 1958. A Practical Manual of Medical and Biological Staining Techniques.New York, Interscience, New York.

Gurr, E. 1958. Methods of Analytical Histology and Histochemistry. London, Leonard Hill.

Gurr, E. (Encyclopaedia of Microscopic Stains). London, Baltimore, William and Wilkins.

Haldane, J.S. and Priestley, J.G. 1935. Respiration. Oxford University Press, London.

Hale, L.J. 1958. Biological Laboratory Data. Methuen, London.

Harrison, G.A. 1947. Chemical Methods in Clinical Medicine. 3rd edn. Churchill, London.

Harrow, B. 1950. Textbook of Biochemistry. 5th edn. W.B. Saunders Co., Philadelphia.

Hawk, P.B., Oser, B.L. and Summerson, W.H. 1965. Practical Physiological Chemistry. Saunders Co., Philadelphia.

Hoar, W.S. 1966. General and Comparative Physiology. Prentice – Hall, Englewood Cliffs, New Jersey.

Horning, E.C., Vanden Henvel, W.J.A. and Creech, B.G. 1963. Methods of Biochemical Analysis. edited D. Click, Interscrew, New York.

Hoyle, G. 1957. Muscular Contraction. Cambridge University Press. London.

Humason, G.L. 1979. Animal Tissue Techniques. 4th edn. W.H. Freeman & Co., San Francisco.

Jackson, J.V. 1960. Chromatographic Techniques. 2nd edn. Vol. I. Edited I. Smith, Heinemann, London.

Jacoby, F. and Martin, B.F. 1949. The Histochemical Test for Alkaline Phosphatase. Nature, 163, 875-876.

King, J. 1965. Practical Clinical Enzymology. Van Nostrand, London.

Kleiner, I.S. 1951. Human Biochemistry. 3rd edn. The C.V. Mosby, Co., St. Louis.

Koch, F.C and Hanke, M.E. 1953. Practical Methods in Biochemistry. 6th edn. William & Wilkins, Baltimore.

Koelle, G.B. and Friedenwald, J.S. 1949. A Histochemical Method for Localizing Cholinesterase Activity. Proc. Soc. Exptl. Biol. Med. 70, 617-622.

Kraijian, A.A. and Gradwohl, R.B.H. 1952. Histopathological Techniques. St. Louis, C.V. Mosby.

Kuksir, A. 1966. Methods of Biochemical Analysis. Edited D. Glick, Interscience, Vol. 14. New York.

Kurmick, N.B. 1949. Methyl green – pyronin. 1. Basis of selective staining of nucleic acids. J. Gen. Physiol. 33, 243-264.

Lederer, M. 1955. Introduction to Paper Electrophoresis. Elsevier, Amsterdam.

Lillie R.D. 1948. Histopathological Techniques. Balkiston Co.

Lillie, R.D. 1954. Histopathological Techniques and Practical Histochemistry. 3rd edn. McGraw Hill.

Mallory, F.B. 1939. Pathological Technique. Saunders, Philadelphia.

Mallory, F.B. 1944. Pathological Technique. Saunders, Philadelphia.

Mathews, A.P. 1939. Physiological Chemistry. 6th edn. William & Wilkins Co., Baltimore.

Mcelroy, W.D. 1964. Cell Physiology and Biochemistry. Engrewood Cliffs, N.J. Prentice Hall Inc.

McLung, C.E. 1939. Handbook of Microscopical Technique. Hoeber. New York.

Mitchel, F. 1965. A Textbook of General Physiology. McGraw Hill, New York.

Mukherjee, K.L. 1989. Medical Laboratory Technique: A Procedure Manual for Routine Diagnostic Tests, Vol. I., Tata McGraw-Hill Publishing Company Limited, New Delhi.

Mukherjee, K.L. 1989. Medical Laboratory Technique, A Procedure Manual for Routine Diagnostic Tests, Vol. III. Tata McGraw-Hill Publishing Company Limited, New Delhi.

Natelson, S. Microtechniques of Clinical Chemistry for the Routine Laboratory, C.C. Thomas, Springfield, Illinois.

Pantin, C.F.A. 1946. Notes on Microscopical Techniques for Zoologists. Cambridge University Press.

Pearse, A.G.E. 1960. Histochemistry: Theoretical and Applied. 2nd edn. Churchill, London.

Pearse, A.G.E. 1968. Histochemistry: Theoretical and Applied. Vol. I. 3rd edn. Churchill, London..

Pearse, A.G.E. 1972. Histochemistry: Theoretical and Applied. Vol. II. 3rd edn. Edinburgh, Churchill, Livingston.

Pigman, W.W. and Goepp, R.M. Jr. 1948. Chemistry of Carbohydrates. Academic Press, New York.

Plummer, D.T. 1989. An Introduction to Practical Biochemistry. 3rd edn. Tata McGraw-Hill Publishing Company Limited, New Delhi.

Prosser, C.L. and Brown, F.A. 1961. Comparative Animal Physiology. Saunders, Philadelphia.

Samson, W. 1971. Applied Physiology. Oxford University Press.

Sayhun, M. 1944. Outline of Amino Acids and Proteins. Reinhold Publishing Corp.

Schmidt, C.L.A. 1944. The Chemistry of Amino Acids and Proteins. 2nd edn. Charles & Thomas, Springfield, Ill.

Singh, S.P. 1990. Manual of Practical Biochemistry. C.B.S. Publishers & Distributors, New Delhi.

Smith, I. 1960a. Chromatographic Techniques. 2nd edn. Heinemann, London.

Smith, I. 1960b. Chromatographic and Electrophoretic Techniques. Vol. I. Heinemann, London.

Spector, W.S. 1961. Handbook of Biological Data. Saunders, Philadelphia.

Stahl, E. 1964. Thin Layer Chromatography. A Laboratory handbook, Academic Press, New York.

Summer, J.B. and Somers, G.F. 1953. Chemistry and Methods of Enzymes. 3rd edn., Academic Press, New York.

Swarup, H., Arora, S. and Pathak, S.C. 1986. Laboratory Techniques in Modern Biology. Kalyani Publishers, New Delhi.

Szent-Gyorgyi, A.G. 1960. Chemistry of Muscular Contraction. Academic Press, New York.

Tuttle, W.W. and Byron, A.S. 1969. Textbook of Physiology. Mosby, St. Louis.

Unbriet, W.H., Burris, R.H. and Stauffer, J.F. 1964. Micrometric Techniques. Burgess, Minneapolis.

Verma, P.S. & Srivastava, P.C. 1994. Advanced Practical Zoology. S. Chand & Company Ltd., New Delhi.

Vogel, A.I. 1961. Textbook of Quantitative Analysis. 3rd edn. Longmans, London.

West, E.S. and Todd, W.R. 1961. Textbook of Biochemistry. Macmillan, New York.

White, A., handler, P. and Smith, E.L. 1964. Principles of Biochemistry. (McGraw Hill, New York), Kogakusha Co., Tokyo.

Williams, R.J. and Beerstecher, E. Jr. 1948. An Introduction to Biochemistry. 2nd edn., D. Van Nostravard Co., New York.

Wintrobe, M.M. 1951. Principles and Applications of Electrophoresis. Elsevier, Amsterdam.

Wood, D.W. 1968. Principles of Animal Physiology. Edward Arnold Ltd.

Woolf, L.I. 1963. Advances in Clinical Chemistry. edited Sobotka, H. and Stewart, C.P., Academic Press, New York.

Wunderby, C.H. 1963. Principles and Applications of Electrophoresis. Elsevier, Amsterdam.